D0906355

Earth Science
Discovering the secrets of the earth

EARTHQUAKES AND VOLCANOES

Grolier Educational

Grolier Educational

First published in the United States in 2000 by Grolier Educational, Sherman Turnpike, Danbury, CT 06816

Reprinted 2001

Author
Brian Knapp, BSc, PhD

Art Director
Duncan McCrae, BSc

Editors
Mary Sanders, BSc and Gillian Gatehouse

Illustrations
David Woodroffe, Julian Baker and David Hardy

Designed and produced by
EARTHSCAPE EDITIONS

Reproduced in Malaysia by
Global Colour

Printed in Hong Kong by
Wing King Tong Company Ltd

Acknowledgments
The publishers would like to thank the following for their kind help and advice: Disaster Prevention Research Institute, Kyoto University; NASA and The United States Geological Survey.

Library of Congress Cataloging-in-Publication Data

Earth science
    p. cm.
    Includes index.
    Contents: v. 1. Minerals — v. 2. Rocks — v. 3. Fossils — v. 4. Earhquakes and volcanoes — v. 5. Plate tectonics — v. 6. Landforms — v. 7. Geological time — v. 8. The earth's resources.
    ISBN 0–7172–7499–3 (set: alk. paper) — ISBN 0–7172–9493–5 (v. 1: alk. paper) — ISBN 0–7172–9494–3 (v. 2: alk. paper) — ISBN 0–7172–9495–1 (v. 3: alk. paper) — ISBN 0–7172–9496–X (v. 4: alk. paper) — ISBN 0–7172–9497–8 (v. 5: alk. paper) — ISBN 0–7172–9498–6 (v. 6: alk. paper) — ISBN 0–7172–9499–4 (v. 7: alk. paper) — ISBN 0–7172–9502–8 (v. 8: alk. paper)
    1. Earth science—Juvenile literature. [1. Earth science.] I. Grolier Educational Corporation.
QE29.E27 2000
550 — dc21

99–086995
CIP

Picture credits
All photographs are from the Earthscape Editions photolibrary except the following:
(c=center t=top b=bottom l=left r=right)
Disaster Prevention Research Institute, Kyoto University 5 (all photographs); NASA 27b, 29b, 43tr, 53b, 58tr; Keith Ronnholm 54–55; USGS 4b, 9b, 10tl, 13t, 14b, 16tr, 16c, 16bl, 17tr, 23cr, 23bl, 24b, 25tl, 25b, 29bl inset, 51bl (Casadevall, T.J.), 9tr (Gilbert, G.K.), COVER, 35br, 37bl, 42bl, 48–49, 49t (Griggs, J.D.), 57cr (Heliker, C.C.), 1, 36br, 52b, 55tl (Austin Post), 47b (Swanson, D.A.), 57b (Tilling, R.I.), 28br (Wallace, RE).

This product is manufactured from sustainable managed forests. For every tree cut down, at least one more is planted.

(Previous page) Plinian eruption from Mount Saint Helens during the 1980 eruption.

(This page) Mount Saint Helens after the eruption. The top of the cone has been blown away, and a large crater now occupies the upper part of the north-facing flank.

Contents

Chapter 1: Earthquakes

As you read this book, somewhere around the world rocks will snap and cause an earthquake, while somewhere else a volcano will erupt. Both earthquakes and volcanoes are the clearest signs that the earth on which we live is not stable but constantly under stresses unimaginably huge.

Earthquakes and volcanoes are hard to predict because they both happen when unseen forces deep within the crust overcome the equally unseen resistance of the rocks to remain unbroken. This unpredictablility is why earthquakes and volcanoes can be the causes of both disasters and misery. But they also tell us of profound changes taking place within the earth's crust. And that is what this book is about: the nature of an earthquake and a volcanic eruption, the landforms each produces, and how both of these features can be explained as part of our understanding of the way the earth works.

(Above and left) The ground often splits apart during an earthquake, although the fissure is not deep. It is just one of many surface expressions of something occurring tens of kilometers below the surface.

By understanding how earthquakes and eruptions occur, it has also been possible to explain much about the landscape around us and to look for signs that new eruptions or earthquakes may be about to happen. Thus, armed with all of this information, earth scientists have also been able to help people understand the kinds of hazards that earthquakes and eruptions pose, how best to predict them, and how to protect against their devastating effects.

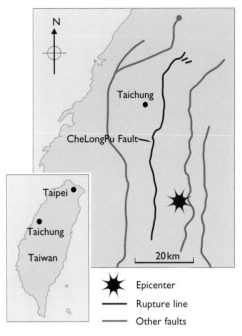

(On this page) At about 1:47 A.M. on the morning of September 20, 1999, an earthquake struck the Pacific island of Taiwan. Within a few moments major changes in the landscape had occurred.

(From top to bottom) One part of the island sank lower than the rest by several meters. The change in height that occurred (along the fault) was seen across towns and countryside. Where the ground sank across the path of a river, it created a new waterfall as well as destroying a bridge.

Where the rocks were unstable or the soil loose, shaking by the earthquake caused rockfalls and landslides. All of these changes had occurred within a few seconds and without warning.

Seismic waves

Because the earth's crust is brittle, when rocks near the surface are put under pressure, they do not bend; they break. An earthquake is the pattern of vibrations in the earth—called SEISMIC WAVES—that occur when rocks break and move to a new position.

The pattern of seismic waves moves out from the place where the rocks break in the same way as waves ripple outward when a pebble is dropped into a pond of still water.

Waves on the surface (which often shake the ground up and down and from side to side very violently) are merely the most obvious part of what happens. Quite unseen, the shock also sends waves through the earth.

The starting point for the break is called the FOCUS, or HYPOCENTER. The position on the ground surface directly above this starting point is called the EPICENTER. The line, or more accurately, the zone, along which the break occurs is called the FRACTURE ZONE, or FAULT.

(Below) The focus (hypocenter) and epicenter of an earthquake.

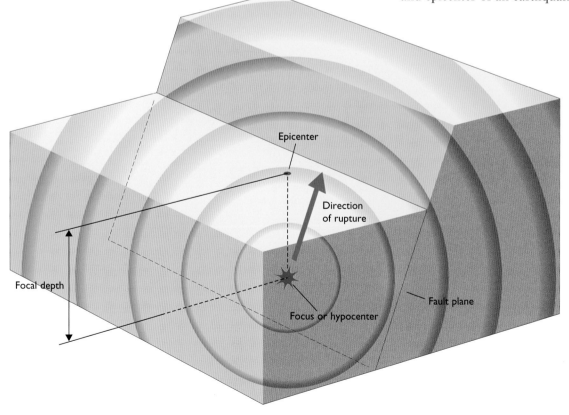

Epicenter

Direction of rupture

Focal depth

Focus or hypocenter

Fault plane

The break begins at a point and then runs along the fracture zone like a piece of cloth being torn. It is called a RUPTURE.

Because the rupture occurs progressively along a line, seismic waves are produced by different parts of the rupture a few seconds apart. The waves produced by the rupture at one point may overlap those produced at another point on the rupture, and this often results in a very complicated pattern of ground movement.

Earthquake waves are normally described in terms of the time it takes for each wave crest to pass. This is called the wave period. Some waves have crests that pass quickly enough for the wave to be audible, so that it is sometimes possible to hear an earthquake as a deep rumble. But the waves that do the damage move more slowly, often with a period of over a second between crests. The slow waves can make the ground pitch and roll; and if this pitching and rolling lasts for long enough, cliffs will crumble, landslides will occur, and buildings will collapse.

(Below) A small fault seen in a roadside quarry. Notice the left-hand part of the white band of rock has been displaced upward compared to that on the right. Although it is inactive now, this fault was, in the past, a rupture zone and the site of repeated earthquakes.

Properties of seismic waves

Earthquakes are measured by an instrument called a seismograph, with the waves being traced on a rotating drum. The trace is called a seismogram.

Seismographs are set up to form networks. The network of seismograms produced can be used to calculate the magnitude of an earthquake, when and where it occurred, the depth of the earthquake focus, and many other properties.

Geologists group the waves produced by an earthquake as those that travel through the rocks, which are called BODY WAVES, and those that travel along the ground, which are called SURFACE WAVES. Body waves provide more information about the earthquake, but surface waves usually have the strongest vibrations and cause most of the damage done by earthquakes.

During an earthquake two types of body waves are produced. One type, known as P WAVES, or primary waves, is similar to sound waves in that they consist of a pattern of pressure waves radiating through the earth. They can travel through any kind of material, liquid or solid.

The second type of waves is known as S WAVES, or secondary waves, which are similar to the waves produced by shaking a rope. They are also called shear waves, and they displace material at right angles to their path. They can only travel through solids. Shear waves travel somewhat more slowly than P waves.

(Below) This is a typical trace, or seismogram. It shows the arrival of the small P and S waves. They are the waves that provide most information about the nature of the rocks through which they pass. A few minutes later the much larger surface waves arrive. They provide limited information about the rocks but are the main cause of property damage, injury, and death. Two types of surface waves are commonly experienced; together they produce a rolling and pitching motion. They are known as LOVE WAVES and RAYLEIGH WAVES.

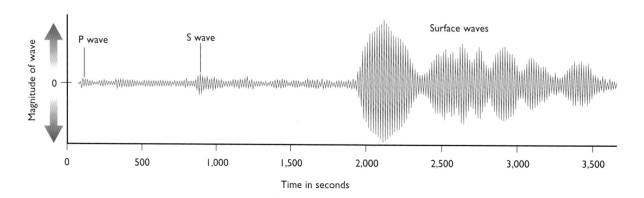

(*Below and right*) As a land wave passes every 10 seconds, the ground ripples along, with waves in solid rock moving past with about ten meters between crests and a third of a meter between trough and crest. Because the land is a solid, the waves make the surface crack up. At a crest gaping fissures occur, while in a trough the rocks are squashed and forced upward in a chaotic way. Then, as the wave passes, crests become troughs, and the fissures close. Troughs become crests, and new fissures open. This can happen for tens of seconds and sometimes for several minutes. The picture above right shows the ground surface after the 1906 San Francisco earthquake.

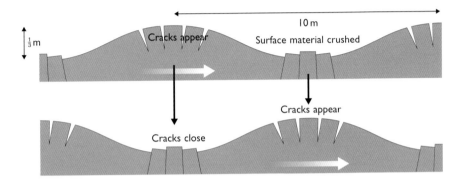

(*Below*) During the 1964 Alaskan earthquake the up-and-down vibrating movement of the surface drove the wooden supporting piles up through the concrete deck of this bridge.

(Left) During the 1964 Alaskan earthquake these railway tracks were torn from their sleepers and refashioned into a wave.

(Below) Railway track before earthquake

(Below) To-and-fro motion can cause a railway line to break.

A seismogram shows P waves arriving first, often as a very short burst of waves felt as a kind of thud or jolt. The S waves follow a few seconds later, giving a kind of swaying feel to the ground. Then some seconds later the surface waves arrive, and they make the ground pitch and roll. The ground can pitch and roll for perhaps half a minute, although some earthquakes go on for much longer.

Earthquake magnitude and intensity

It is possible to work out what the total energy released by the earthquake might have been, while it is difficult to find a single good measurement for an earthquake. The energy released is called the magnitude of the earthquake, measured by the RICHTER SCALE. It is most useful for scientific investigations. But the magnitude does not describe the effect of the earthquake, and for most people it is the intensity of the effect, not the overall amount of energy, that is crucial. The intensity, as expressed by the MODIFIED MERCALLI SCALE, is much more subjective and attempts to give a value to the effect of the earthquake.

The Richter Scale, named after Charles F. Richter of the California Institute of Technology, is logarithmic, meaning that each one-unit increase on

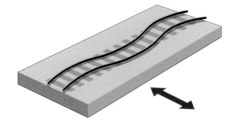

(Below) The sideways movement of S waves can make the ground sway from side to side.

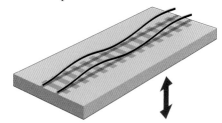

(Below) The effect of rolling surface waves can cause the tracks to buckle up.

(Table below) Comparison between earthquake magnitude (Richter Scale) and intensity (Modified Mercalli Scale).

Magnitude	Intensity	Description
1.0–3.0	I	I. Not felt except by a very few people under especially favorable conditions.
3.0–3.9	II–III	II. Felt only by a few persons at rest, especially on upper floors of buildings. III. Felt quite noticeably by people indoors, especially on upper floors of buildings. Many people do not recognize it as an earthquake. Standing cars may rock slightly. Vibrations similar to the passing of a truck. Duration estimated.
4.0–4.9	IV–V	IV. Felt indoors by many people, outdoors by a few during the day. At night some people awakened. Dishes, windows, doors disturbed; walls make cracking sound. Sensation like heavy truck striking building. Standing cars rocked noticeably. V. Felt by nearly everyone; many awakened. Some dishes, windows broken. Unstable objects overturned. Pendulum clocks may stop.
5.0–5.9	VI–VII	VI. Felt by all; many people will be frightened. Some heavy furniture moved; instances of fallen plaster may occur. Damage slight. VII. Damage negligible in buildings of good design and construction; slight to moderate in well-built ordinary structures; considerable damage in poorly built or badly designed structures; some chimneys broken.
6.0–6.9	VIII–IX	VIII. Damage slight in specially designed structures; considerable damage in ordinary substantial buildings with partial collapse. Damage great in poorly built structures. Fall of chimneys, factory stacks, columns, monuments, walls. Heavy furniture overturned. IX. Damage considerable in specially designed structures; well-designed frame structures thrown out of plumb. Damage great in substantial buildings, with partial collapse. Buildings shifted off foundations.
7.0 and higher	X or higher	X. Some well-built wooden structures destroyed; most masonry and frame structures destroyed with foundations. Rails bent. XI. Few, if any (masonry), structures remain standing. Bridges destroyed. Rails bent greatly. XII. Damage total. Lines of sight and level are distorted. Objects thrown into the air.

the scale represents a tenfold increase in the energy released by the earthquake. Thus an earthquake of magnitude 6 is 10 x 10 x 10, or 1,000 times more energetic than an earthquake of magnitude 3. People cannot feel an earthquake smaller than magnitude 2.

Major earthquakes are those with magnitudes greater than 6, and great earthquakes have magnitudes greater than 8 on the Richter Scale. The Richter Scale is a continuous scale, so that an earthquake can have a magnitude 5.7, 5.8, and so on. There are, on average, 20,000 earthquakes with a magnitude greater than 4 occurring somewhere in the world each day.

The Modified Mercalli Scale expresses the intensity of an earthquake's effects in a given locality in classes ranging from I to XII. This ranges from I—not felt except by a very few people under especially favorable conditions, to XII—damage total.

The Richter Scale relies on seismograms, which is why it is able to pick up the earthquakes worldwide. The Mercalli Scale relies on eyewitness reports and is only used for populated areas. The maximum intensity experienced in the Alaska earthquake of 1964 was X; damage from the San Francisco 1906 and New Madrid 1812 earthquakes reached XI.

The reason why two scales are used is that large magnitude earthquakes do not necessarily cause the most intense surface effects. This is because the effect of an earthquake depends on the period of the waves and the materials they pass through. An area with unstable ground underneath it (sand, clay, or other unconsolidated materials), for example, will shake like a bowl of jelly, amplifying the waves, while an area equally distant from an earthquake's epicenter firm ground such as granite underneath it will experience much smaller effects.

An earthquake's destructiveness depends on many factors. In addition to magnitude and the local geological conditions, they include the focal depth of the earthquake.

The geology of an earthquake

The nature of earthquakes is most easily understood through an example. A relatively recent and very well-documented earthquake occurred in Los Angeles in 1994. It was called the Northridge earthquake.

The Los Angeles area (the San Fernando Valley) is a part of California that geologists know as the Transverse Ranges. The famous San Andreas Fault runs to the east of the city, and the main fault system close to Los Angeles is called the Oak Ridge Fault System.

The Northridge earthquake struck Los Angeles at 4:30 A.M. on January 17, 1994. The epicenter of the earthquake at Northridge, just north of Los Angeles, experienced a shock of magnitude 6.7. The effect of the shock was to cause devastating damage rather than huge loss of life, and so was typical of the way in which earthquakes affect modern industrial cities. Fifty-seven people were killed by the earthquake, and 9,000 were injured, a very small number compared to the area population of over fifteen million people. But at the same time, the amount of destruction was large.

This was only a moderate-sized earthquake whose ground-shaking lasted between 10 and 20 seconds. The main effects were to cause buildings to collapse, highway bridges to fall down, and water, gas, and electricity supplies to be cut.

The initial rupture of the ground that caused the Northridge earthquake lay about 17.5 kilometers below the surface. The rupture then extended itself, like a zipper being undone, growing northwestward at about 3 kilometers per second for 8 seconds. It also spread upward, so that when it stopped, the end of the rupture zone was just over 5 kilometers below the surface.

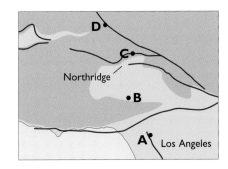

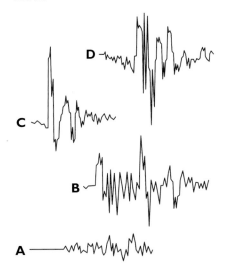

(Above and below) The Northridge earthquake produced quite different kinds of tremors even over a short distance, as the seismograms below show.

(Below) The landscape of the western Los Angeles Basin.

This, like the other earthquakes in the past in this area, occurred on a "blind" thrust fault, meaning that it was a dead-end spur, not part of a continuous fault system.

In fact, major shocks came from three distinct places on the rupture zone, producing overlapping shock waves. The waves were larger in the direction of the rupture, which in this case meant that the most severe waves were sent northwestward out into sparsely populated country.

The time between wave crests for the strongest waves was 1–3 seconds, which are the wave times most liable to cause taller buildings to collapse. But since these waves were directed out into the country where there are no high-rise buildings, the threat to downtown areas was small.

The earthquake affected an area of 4,000 square kilometers, lifting the San Fernando Valley and mountains in a slight dome, with the greatest lifting of about 40 centimeters being where the fault plane comes closest to the surface.

(Below) The shocks associated with the Northridge earthquake are shown on this diagram.

An earthquake is a rupture process that expands from an initial point on a fault plane called the hypocenter. The Northridge earthquake ruptured the fault plane for about 8 seconds and had an average slip across the plane of about 1 meter. The rupture extended across the plane northwesterly from the hypocenter at about 3 km/s. However, some parts of the plane exhibited little or no movement, and some parts slipped more than 3 meters.

The red circles show the locations of the thousands of small aftershocks that followed the initial rupture. The presence of the older faults (blue lines) across the path of the Northridge rupture may have stopped the initial rupture from going any further.

There were a thousand aftershocks each day in the first month following the rupture in January 1994, and even by January 1996 there were still a few each day. Of the aftershocks, nine were greater than magnitude 5.0, and 42 had a magnitude of more than 4.0.

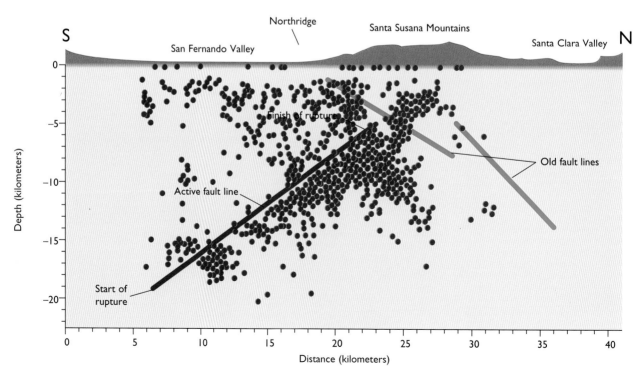

When an earthquake occurs, the stress is relieved in the rocks that move; but places nearby that were not previously under stress may become stressed as a result. They, in turn, may fail, causing further earthquakes. This stress is the reason for the many smaller earthquakes (called AFTERSHOCKS) that are felt after a large earthquake. Aftershocks may occur for many years following a major event.

The pattern of aftershocks that occur within the hours days and months after a large earthquake follows a logarithmic rule: every large magnitude shock will be followed by ten shocks that are a tenth as powerful, a hundred that are a hundredth as powerful, and a thousand that are a thousandth as powerful.

Patterns of earthquakes

It is now possible to monitor the pattern of earthquakes across the world. The maps of earthquakes clearly show that in any month, for example, there are some small and some big earthquakes. The maps also show that there are some places where earthquakes are common, and others where they are rare.

(Below) The pattern of earthquakes recorded over a long period of time clearly picks out the edges of the tectonic plates. Notice that the depth of earthquakes varies across the world. In general, earthquakes are shallow where continents collide or where ocean plates pull apart. But they are much deeper where ocean plates are pushed below thick continents, as, for example, around some parts of the Pacific Ocean.

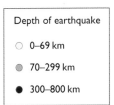

Depth of earthquake

○ 0–69 km

◐ 70–299 km

● 300–800 km

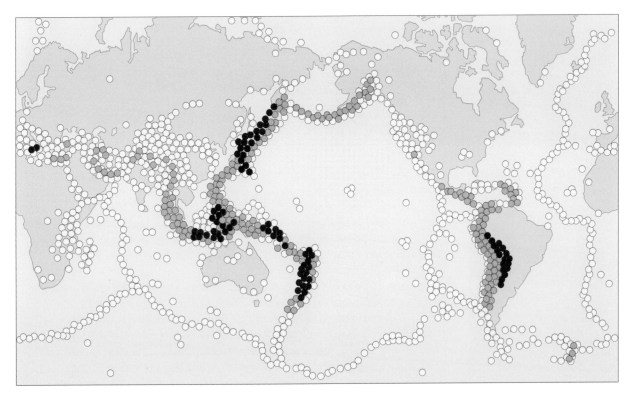

The map on page 15 shows that more large earthquakes occur around the edges of Asia than anywhere else, but that there are also important clusters along the western coast of the Americas. On the European continental mainland there are very few. To the east of Asia the pattern of earthquakes follows a series of curves, or arcs, each bulging to the east.

Scientists now know that the most common

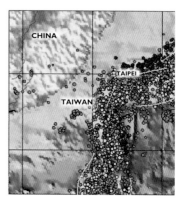

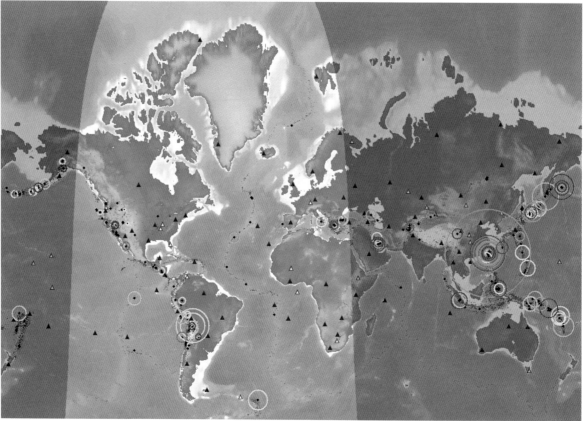

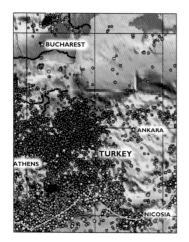

(This page) Earthquakes are happening all of the time. You can now see the earthquakes that are occurring around the world in almost real time on the Internet. This, for example, is an image downloaded from IRIS/USGS soon after the Taiwan earthquake (magnitude 7.6) of September 20, 1999; the largest circle on the map shows the Taiwan earthquake. It also shows the earthquakes that occurred during the previous five years. The circle sizes and colors show the magnitude and the age of the earthquake. This image is re-created every 30 minutes. The dark shadow shows the region of darkness over the earth at the time the image was compiled.

The top map shows the pattern of earthquakes in the Taiwan area. Notice that it follows a curved line. This is an ocean-continent tectonic plate boundary. By contrast, the bottom map shows the pattern of earthquakes around Turkey, a continent-continent plate boundary. There is a more widely scattered pattern here.

places for earthquakes to form are at the boundaries of the dozen or so large slabs of the earth's crust known as the TECTONIC PLATES. (For a full description of plate tectonics see the book *Plate tectonics* in this set.)

Each type of plate boundary proves to have its own distinctive types of earthquakes. At SPREADING BOUNDARIES, where two plates pull apart, earthquakes occur at shallow depths (within 30 kilometers of the surface). In part this is because the crust at a spreading boundary is very thin. The fault lines tend to be nearly vertical.

Where plates grind past one another, the earthquakes also tend to occur at shallow depths and form fairly straight long lines. Here the earthquakes tend to be shallow. The San Andreas Fault is the world's best-known example of a line on which earthquakes are common.

At DESTRUCTIVE PLATE BOUNDARIES, where plates collide, and where one plate pushes the other downward into the mantle, earthquakes occur at a wide range of depths, from shallow to very deep.

Ninety percent of all earthquakes occur at plate boundaries. The rest occur within plates, but they are not insignificant. Often they form along very ancient plate boundaries that have subsequently become part of new plates. The great New Madrid earthquake of 1812 was one of them. Others occur where plumes of mantle material rise up underneath a plate. They are called HOT SPOTS. The area near the northern end of the East African Rift Valley is one of them.

San Francisco

San Andreas Fault

(*Above*) This computer-generated image shows the many faults that lie across the San Francisco Bay region. It is a classic seismic gap, where earthquakes are rare because the ground is locked more firmly together than other places on the San Andreas Fault. When they do occur, the earthquakes are of larger magnitude than elsewhere because they have a long period of catching up to do.

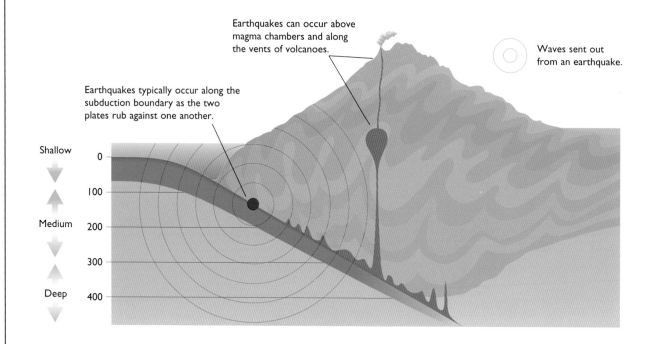

Earthquakes can occur above magma chambers and along the vents of volcanoes.

Waves sent out from an earthquake.

Earthquakes typically occur along the subduction boundary as the two plates rub against one another.

Shallow

0

100

Medium

200

300

Deep

400

Earthquakes occur at greatly varying depths. The depth is often called the FOCAL DEPTH, meaning the distance vertically from the surface to the earthquake focus (see diagram on page 6). Earthquakes with focal depths from the surface down to about 70 kilometers are classified as shallow. Earthquakes with focal depths from 70 to 300 kilometers are classified as intermediate. Earthquakes whose focal depth is greater than this are deep; those below 400 kilometers (the depth of the thickest crust) must occur in the upper MANTLE. Earthquakes that occur as a result of plates colliding (as shown in the diagram above) can be shallow, intermediate, or deep. The mountains around the Pacific Ocean are on this kind of boundary.

Seismic gaps

Earthquakes are part of long-term movement of the crust. Stresses do not often build up to be very big. In many places they are relieved by the multitudes of small earthquakes and aftershocks that happen every day. Through these small, often unnoticed quakes blocks of land creep past one another, and the tectonic plates drift over the earth's surface.

(Above) This diagram shows a place where two plates meet, forcing one into the earth. It is called a SUBDUCTION ZONE. Earthquakes are generated along the whole face, where the two slabs scrape against one another. As a result, shallow, intermediate, and deep earthquakes are produced. This explains why subduction zones (such as those that occur along the western side of the Pacific Ocean) are the most earthquake-prone places on earth.

A large earthquake often represents a stress that has been building up in the crust for a long time without being relieved. In effect it is a place where rocks have become jammed and unable to move. A part of a long fault where no earthquakes have occurred for a long time is called a SEISMIC GAP.

When the rocks do finally shift in a seismic gap, much more movement occurs than normal as the rocks "catch up" on movements that have already occurred in the fault nearby.

For example, the total movements from earthquakes and creep along the San Andreas Fault are over 500 kilometers in the last twenty million years. The average rate at present is about four to five centimeters each year. But in some places the fault has become stuck and only moves occasionally; and when it does move, it does so very violently. One of these seismic gaps is in the area of San Francisco, which is why this city suffered only two earthquakes in the 20th century, but why both were large.

(Below) All parts of a plate boundary move on average at the same rate. However, there are some parts of the plate boundaries where earthquakes are infrequent. They are known as seismic gaps. Their locations are shown on this map. When earthquakes do finally occur in seismic gaps, they are likely to be more powerful than along parts of the plate boundaries where earthquakes happen frequently. In this way a few large earthquakes will allow the seismic gap area to catch up with the more frequent movements elsewhere.

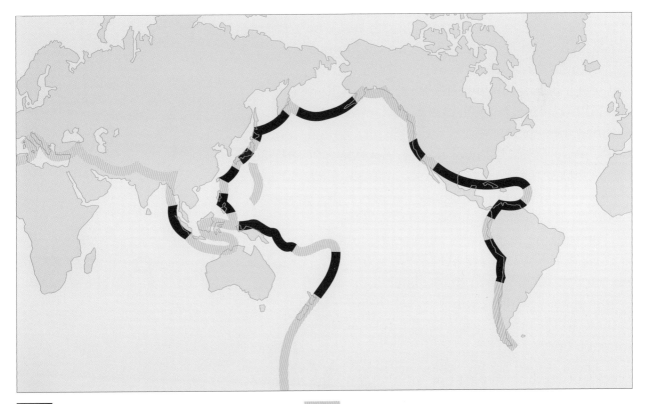

■ Seismic gap where stresses build   Frequent earthquakes relieve stress

Making use of earthquake waves

Earthquakes are usually thought of as being destructive events, but they can produce a large amount of information that is of great value to earth science. Not only can they locate the edges of the earth's plates and tell us about the way in which plates are dipping back into the earth, but they can also tell us about the whole nature of the earth. They are, in effect, a geologist's stethoscope to the earth's mantle and CORE.

To understand how earthquakes can be used in this way, we need to concentrate not on the waves that reach the surface, but on those that go down into the earth.

When an earthquake occurs, the waves are detected by monitoring stations all over the world, not just by those close to where the earthquake occurs. Detection is possible worldwide because seismic waves move right through the earth.

By studying the time it takes for earthquakes to reach distant parts of the planet, it is possible to work out what the rocks are like between where the earthquake occurred and where it was monitored.

Using the nature of material in the earth's crust as a starting point, it is possible to work out how long it should take an earthquake to be detected at distant stations. In general, waves move faster through more dense material and slower through less dense material. Observations showed that waves arrived much earlier than expected. This could only happen if the interior of the earth was more dense than the crust.

By looking at patterns of P waves and S waves, it is clear that S waves are absent from traces recorded on the side of the earth immediately opposite from the earthquake. S waves cannot travel through liquids, leading to the conclusion that the core of the earth must be liquid.

There is also a zone where neither P nor S waves are recorded. It is known as a SHADOW ZONE. It can only occur if the mantle and core bend the P waves as

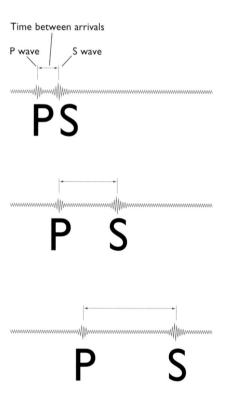

(Below) Because P waves travel faster than S waves, the time interval between the arrival of the P waves and the S waves increases away from the focus. Studying traces such as this allows scientists to calculate the distance to the earthquake as well as the structure of the earth.

Time between arrivals

P wave | S wave

PS

P S

P S

the P waves pass through the zone. It is possible to use this information to work out the size and the nature of the various layers inside the earth.

In this way, from earthquake traces alone, it has been possible to find out the thickness of both the core and the mantle, and to find out what they are made of.

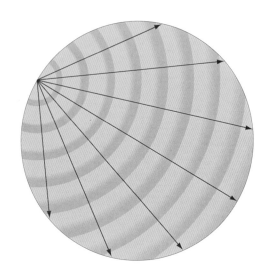

(Right) This diagram shows the way in which waves would spread out if the earth was made of exactly the same material right to the core. Notice that the waves move in straight lines.

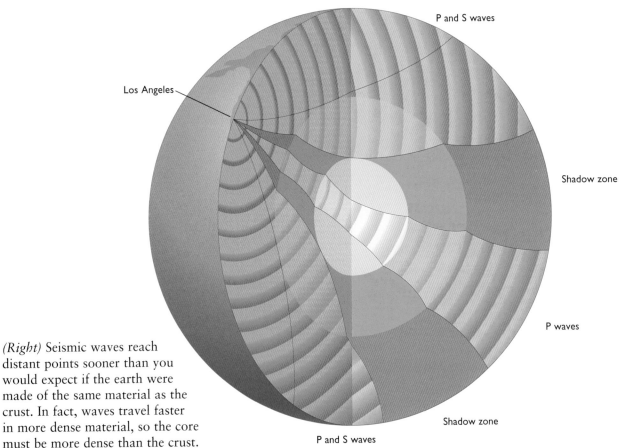

P and S waves

Los Angeles

Shadow zone

P waves

Shadow zone

P and S waves

(Right) Seismic waves reach distant points sooner than you would expect if the earth were made of the same material as the crust. In fact, waves travel faster in more dense material, so the core must be more dense than the crust.

Chapter 2: Landforms and earthquakes

In the previous chapter we saw how earthquakes can produce a variety of waves, and that they can be helpful in providing information about the earth's great tectonic plates and even about the nature of the center of the earth. In this chapter we look at how earthquakes alter the landscape. To do this, we will first look at the example of the Alaskan earthquake of 1964. Then we will look at how repeated earthquakes can give rise to some dramatic landscapes.

The Alaskan earthquake

The great Alaskan earthquake, which lasted for about three minutes, was felt over a large area of Alaska and in parts of western Yukon Territory and British Columbia, Canada. It occurred in a region of the world with quite a small population; so although 125 people were killed (110 from the TSUNAMI—often called a tidal wave—15 from the earthquake), this was very low by most earthquake standards. For the same reason, damage costs were also quite low.

The epicenter was under Prince William Sound (now famous for the Valdez oil spill), but many of the most serious effects occurred elsewhere. For example, at Anchorage, about 120 kilometers northwest of the epicenter, there were large surface vibrations and landsliding.

LANDSLIDES affected many properties that had been built on soft rock, including the downtown business section.

A huge area—520,000 square kilometers—around the earthquake focus was moved vertically. About half of the area was lifted up; the rest sank lower. The major area of uplift extended northeast from southern Kodiak Island to Prince William Sound. The biggest uplift was 15 meters, while the biggest drop was 2.3 meters.

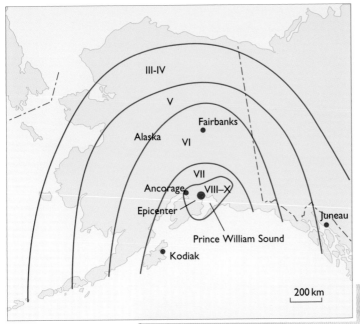

(Left) The shape of Alaska and the location of the 1964 earthquake near the largest settlement, Anchorage. The map shows the center of the earthquake and the various zones that were affected by it. The numbers refer to earthquake intensity (how much damage was done). The intensity scale is given on page 11.

III-IV

V

Alaska

Fairbanks

VI

VII

Ancorage VIII–X

Epicenter

Prince William Sound

Kodiak

Juneau

200 km

(Above) Closeup of Government Hill elementary school, which was destroyed by the Government Hill landslide at Anchorage, Alaska.

(Left) Houses collapsed in a zone of rifting.

23

This shock generated a tsunami that devastated many towns along the Gulf of Alaska and was felt as far away as Hawaii. The maximum wave height recorded was 67 meters at Valdez Inlet.

Landscape effects

Although most people noticed the landsliding and the collapse of buildings, by far the biggest effects on the landscape were the changes in height that occurred. It became very obvious when places that were previously at sea level suddenly became dry land.

While the whole region was domed up, old fault lines of weakness in the rocks were exploited by the earthquake, so that the landscape became stepped in places.

A single break in the landscape produces a stepped profile; but in the Alaskan earthquake breaks occurred in parallel, so that whole sections of the land dropped down between others. This is known as RIFTING.

(Below) The very gently sloping rocky platform at Cape Cleare, Montague Island, Prince William Sound, that had been eroded by wave action over the centuries, was lifted some 10 meters during the earthquake, exposing about 200 meters width of new land. The white coating is made up of the shells of marine life that died when their seabed home became dry land.

(Left) Rifting occurred at the head of L Street in Anchorage during the earthquake. Land sank about three meters as a result of the area to the left moving nearly four meters sideways.

A number of houses seen in this photograph were undercut or tilted by subsidence of the rift.

(Above) This picture looks southwest along the Hanning Bay Fault on southwest Montague Island in Prince William Sound. Other earthquakes had occurred along a known fault—an established weakness in the crust. It was not visible before the earthquake, but the ground to the right of the fault (the straight line running across the foreshore) was lifted up five meters compared to the left-hand side. In addition, the *whole* area was also lifted above sea level. As a result, the fault became marked by a step across dry land in an area that was previously under the sea. It is believed that this fault is nearly vertical.

Landscapes of faults and rifts

Faults are surfaces (planes) of weakness in the earth's crust that are used time after time by earthquakes. They occur because the earth's plates are always moving. This puts great stress on the brittle crust, and so it often breaks up. Each break is a fault. Faults are a common feature of the landscape.

There are three kinds of fault, depending on how the blocks of crust move. If the blocks pull apart, one block will slide down the fault plane. It will leave part of the fault plane exposed as a sloping surface, often looking like a cliff. It is called a NORMAL FAULT. Normal faults usually have steep fault angles so that although one block may drop a considerable height, the horizontal distance it moves will be small.

If the crust is squashed for example, when two plates push against one another then because one piece of crust is being pushed over another, and the crust is shortening, the faults are called THRUST FAULTS or REVERSED FAULTS. Most thrust faults have fault planes with very low angles, so that blocks are pushed over one another for long distances, sometimes for tens of kilometers. They are common only where mountain ranges are forming.

If two slabs of rock are not pulling apart or colliding, but simply scraping past one another, the rocks along the zone of scraping will be broken up into a jumble of pieces. The faults produced will not make the land rise or fall, but simply shift it sideways. These are called LATERAL FAULTS, TRANSFORM FAULTS or TRANSCURRENT FAULTS. Lateral faults are often very long as, for example, in the case of the San Andreas Fault, which extends for thousands of kilometers along the west coast of the United States.

Faults rarely occur singly. Usually a brittle, weak part of the crust will split into many fragments, and so a large number of faults will be found together. In places where the crust is stretching, many parallel faults will be produced, and here two faults may

(Below) A normal fault usually has a steep fault angle, so that the majority of the movement is vertical.

(Below) A reversed or thrust fault pushes one block over another during the shortening of the crust. It has a shallow angle.

(Below) A transcurrent fault makes one slab scrape past another. It has a vertical angle. There is no significant vertical change in height between slabs. The diagram shows a river whose course reflects the movement of the fault.

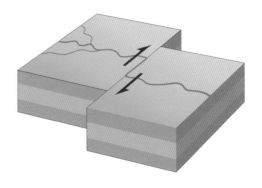

allow the block in between to sink down. This results in a **RIFT VALLEY** or **GRABEN**.

If rift valleys develop on both sides of a block that is left standing up, the block is known as a **HORST**.

Many rift valleys have much the same width, often in the range of 30 to 45 kilometers. This may be the largest span that the crust can support before it collapses.

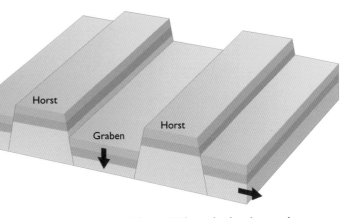

(Above) When the land stretches, blocks of the crust fracture along parallel lines, and then some blocks sink, or founder, between others. The foundering blocks make rift valleys, while the blocks that remain upstanding make block mountains or horsts.

Rift valley landscapes

There are some spectacular examples of rifting, such as the Basin and Range region of the southwestern United States. Here the crust has been gently doming up and stretching for a long time, making the crust sixty kilometers wider in the last fifteen million years, and allowing blocks to sink and create rift valleys. They are the basins. The horsts that remain standing are the ranges.

One of the best-known rift valley basins in the world is Death Valley, whose floor is 83 meters below

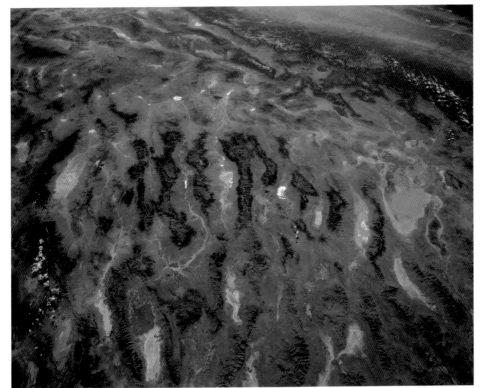

(Left) This satellite picture of the Basin and Range region of the southwestern United States shows parallel dark-colored horsts (ranges) separated by white-colored basins. The white of the basins is produced by salt lakes. The view looks southwestward across Nevada to the snow-capped peaks of the Sierra Nevada mountains of California.

sea level. The nearby Panamint Range is nearly 4,000 meters above sea level. Erosion has stripped away some of the mountains, and this material has partly filled in the valley, so the true surface of the solid rock floor in Death Valley is many hundreds of meters lower still. The steep face leading from the top of the Panamint Range to the rock floor of Death Valley is a normal fault, the result of numerous small shifts of height, each one accompanied by an earthquake.

The whole of East Africa has been lifted up into an even bigger dome, and long rift valleys have been formed on either side of the dome. This has produced the East African Rift Valley system, and in it lie some of the world's biggest lakes.

Transcurrent fault landscapes

When crust moves sideways as opposed to vertically, there is less obvious impact on the landscape. This is the case with the San Andreas Fault in California, for example, which is not readily obvious from the ground. This fault moves, on average, five centimeters

(Above) Death Valley is part of the Basin and Range region of southwestern United States (see also the satellite picture on page 27). Death Valley, with the white covering of salt, is the rift, or graben, while the mountains to either side are horsts. This picture shows the Panamint Range.

(Below) Part of the San Andreas Fault.

28

each year. And so, even though it does not have a major impact on the landscape, it is a source of frequent earthquakes and causes considerable damage. The landscape effects are best seen from the air, where the shattered rocks along the fault zone can be seen. Other evidence includes river courses that have been displaced by the fault. During the 1906 earthquake in the San Francisco region, roads, fences, and rows of trees and bushes that crossed the fault were offset by several meters. In each case the ground to the west of the fault moved northward.

(Below) The San Andreas Fault in California is the world's biggest transcurrent fault. It runs through the San Francisco region and can be clearly seen as a line of lakes just south of the Bay Area. This astronaut's view looks northeastward across San Francisco.

(Inset) A computer-generated 3-D model of part of the San Francisco area. The red lines show faults.

San Francisco

San Francisco Bay

San Francisco

San Andreas Fault

Chapter 3: Volcanoes

The earth is believed to have formed when gravity pulled the dust of space into a planet. During this process an incredible amount of heat was released.

The heat contained in the early earth could only be lost in three ways: by CONDUCTION, by RADIATION, or by CONVECTION. The early earth did not have a crust to prevent heat being lost to space, and so most heat was probably lost by conduction and radiation, and new heat was probably brought to the surface by convection. But with time the surface cooled, and a crust formed.

Rocks are not good conductors of heat, so the heat produced inside the earth simply could no longer be lost through the earth's crust. As a result, very little heat could be lost by radiation or conduction. Thus, as the crust developed, the rate at which heat was lost must have slowed down. The core of the earth, however, remained molten and very much hotter than the crust, providing the conditions for convection to continue.

(Above) The early earth probably lost heat by radiation, conduction, and convection.

Convection: the source of volcanoes

Just as the water in a saucepan churns when it is heated from below, so the heat stored within the earth can cause a slow circulation to occur in the mantle and bring hot rock close to the surface. Earth scientists think that the rock in the outer mantle may be hot enough to move slowly and even may be molten in places.

Comparing the upper mantle with the way in which heated water moves in a saucepan helps us understand the way in which heated rock may move just below the crust. In a saucepan hot water currents tend to rise in columns before sinking again. This circulatatory movement makes up a cell. Many cells

(Above) As the earth's surface cooled, a crust formed, preventing conduction and radiation.

may form in a heated saucepan of water; if convection inside the earth happens in a similar way, we would expect the pattern of convection also to produce cells, with molten rock rising in some places and sinking down in others. Of course, the pattern of convection in a spherical earth would be more complicated than in a saucepan. Furthermore, the material in the mantle would move extremely slowly.

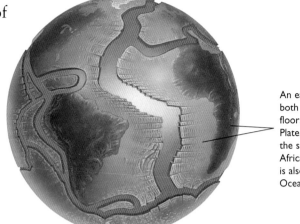

An earth plate contains both continent and ocean floor. This is the African Plate, the part of it above the sea we know as Africa. The African Plate is also part of the Atlantic Ocean floor.

Convection and magma

If convection can work in the outer zones of the earth, this has immense implications for the way in which the crust moves. The earth's crust is very thin compared with the thickness of the mantle in which the convection currents may work. So, any convection currents operating in this region would be able to apply an immense force to the underside of the tectonic plate. Convection could therefore be expected to be powerful enough to make the crust move.

(*Above*) With the crust formed on the surface, convection breaks up the crust, allowing new magma to well up from the mantle along some boundaries.

Where the material rises up, the crust may well be split apart, allowing molten material (MAGMA) to flow to the surface and form volcanoes. Where material sinks again, it may well draw the crust together before trying to pull it back into the mantle. The crust, is, however, much less dense than the mantle and so would not readily be drawn down. Rather, the edges of the tectonic plates would be likely to collide and crush the crust to form mountain ranges. If, however, one piece of plate were to be dragged back down into the mantle, as it descended, it would be likely to melt, releasing low density material that could then slowly rise back through the crust and erupt as volcanoes.

There is much evidence to support this idea, and it does help explain many features of volcanic activity.

Volcanoes, for example, are common along the boundaries of most tectonic plates. Furthermore, the world has essentially two types of volcano, one kind (with runny lava and little explosive activity) forming where the crust appears to be splitting apart, and the other kind (with sticky lava and explosive violence) forming where two plates collide.

The nature of volcanic activity

Trying to understand volcanoes is difficult because of the complex way in which they behave. A volcanic eruption may involve rivers of lava or violent explosions. The same volcano may, on one occasion, erupt lava and, the next time, send up towering clouds of ash and gas and no lava, or be explosive at the start and change to producing lava at the end.

Volcanoes are very unpredictable not just in the materials they spit out, but in how often they erupt. They may erupt every few days, every few years, or

Oceanic rift

Lava flows

Spreading seafloor

Spreading seafloor

Dikes

(Above) At spreading boundaries the splitting of the crust provides an opportunity for magma to well up to the surface and spread out from fissures rather than from single vents.

(Below) The convection cells in the upper part of the earth could be the cause of crustal movement. Volcanoes appear near both spreading and colliding boundaries.

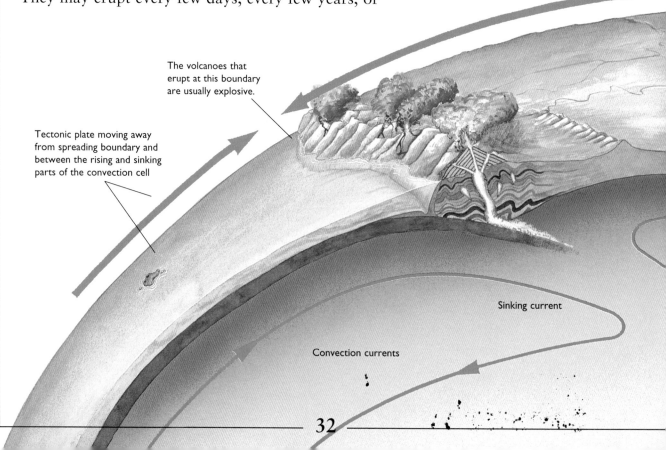

The volcanoes that erupt at this boundary are usually explosive.

Tectonic plate moving away from spreading boundary and between the rising and sinking parts of the convection cell

Sinking current

Convection currents

(Right) Where plates collide and one plate sinks back into the mantle, magma is released that can melt its way back to the surface and provide the source material for volcanoes. Most volcanoes in this location erupt violently and produce large amounts of ash as well as some lava.

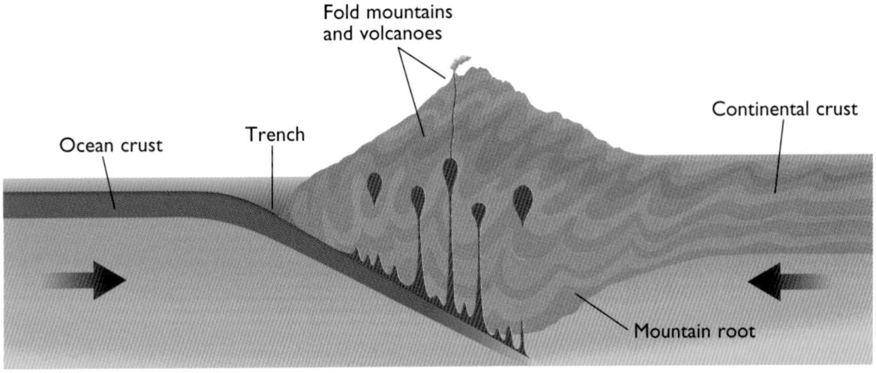

Ocean crust

Trench

Fold mountains and volcanoes

Continental crust

Mountain root

(Right) The distribution of volcanoes that have been active in the last three centuries and their association with spreading plate boundaries (purple lines), destructive or colliding plate boundaries (pink lines), and hot spots.

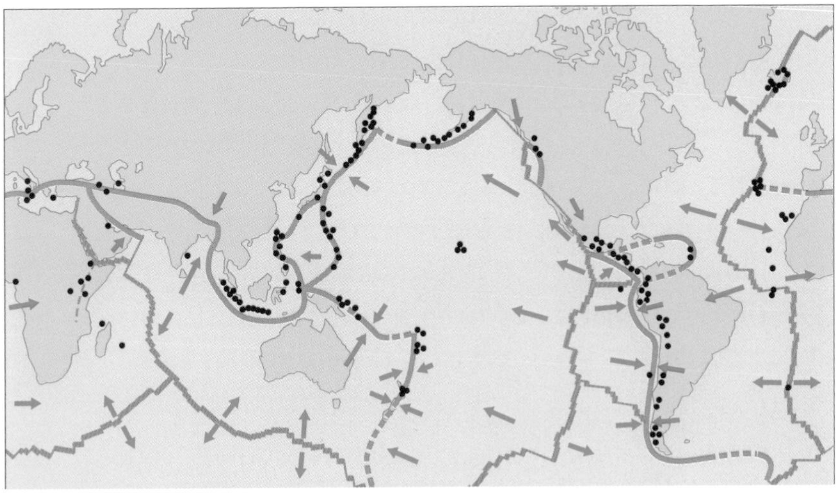

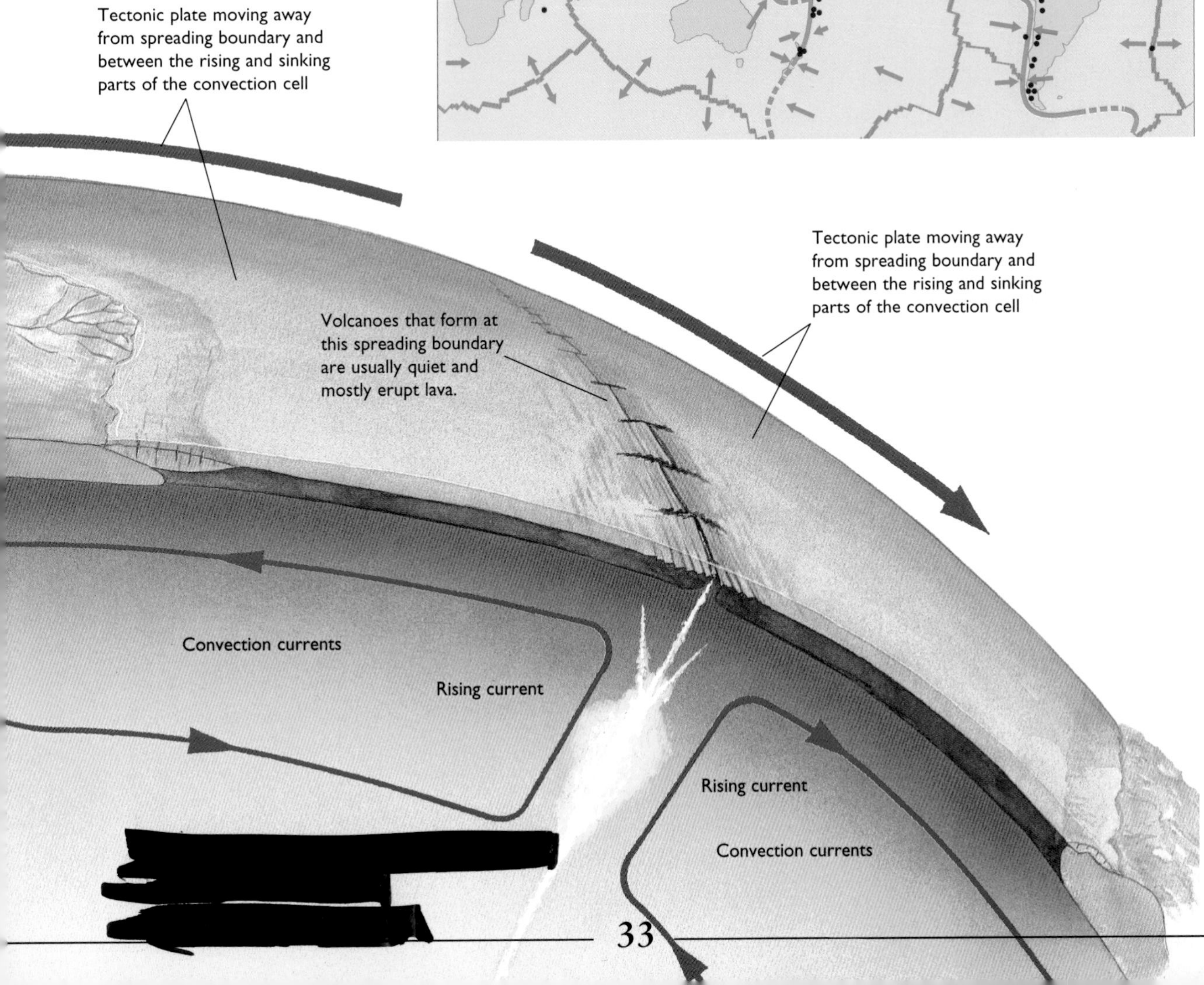

Tectonic plate moving away from spreading boundary and between the rising and sinking parts of the convection cell

Tectonic plate moving away from spreading boundary and between the rising and sinking parts of the convection cell

Volcanoes that form at this spreading boundary are usually quiet and mostly erupt lava.

Convection currents

Rising current

Rising current

Convection currents

every few centuries. This not only makes it difficult to know how the volcano will behave in the future, but also whether it has become inactive. Volcanoes that have erupted within the last ten or fifteen years are generally called ACTIVE VOLCANOES; those that have not erupted for decades, but whose lava and ash is still relatively fresh, are thought of as DORMANT VOLCANOES; and only those with no historic record are regarded as EXTINCT VOLCANOES.

Volcanoes are unpredictable because, like the stress that builds up and suddenly causes an earthquake, stress builds up inside the earth with no sign of change occurring at the surface. How much pressure is needed to cause renewed activity is not easy to guess; neither is the rate at which the pressure is building. An eruption is a sudden release of pressure from within or from below the crust. Once the pressure has been released, the eruption stops.

Magma

Whatever the kind of material that is erupted from a volcano, it has a common source—a chamber deep underground, usually in the lower part of the crust. It is called a MAGMA CHAMBER.

Magma is a mixture of molten materials and liquefied gases. It can be made up of a variety of

(Below) A magma chamber has its source deep within the crust, often in the upper mantle. A large plume of molten rock begins to rise, melting its way upward over perhaps millions of years. When it is close to the surface, the upper part of the crust begins to break up, and this allows magma to flow to the surface as a volcano. A magma chamber may feed a number of volcanoes. When it eventually cools, a magma chamber solidifies into granite and is called a BATHOLITH. Batholiths are found at the cores of many mountain systems.

materials, which is partly responsible for the different types and frequency of eruptions.

The most important control is the proportion of silica to other parts (mainly iron and magnesium compounds) of the magma. If the proportion of silica (which will form quartz when it cools) is above about 60 percent, the magma will be very sticky and will almost certainly erupt explosively. If the magma contains between about 45 and 60 percent silica, it is of intermediate nature and will probably erupt with medium violence. If the lava has less than about 45 percent silica, then it will probably erupt quietly as a fountain of liquid rock with almost no explosive activity.

Types of eruptive material

Volcanoes eject a wide variety of types of materials and thus have an equally wide variety of ways of erupting. This affects both the area in which they erupt and the shape of the volcano that builds up.

Eruptions of liquid magma—called LAVA—are associated with magma that is runny and has a low gas content.

When sticky magma is full of gases, it boils explosively as it rises to the surface and blasts the magma into tiny droplets. In this case the droplets cool as they fly through the air and fall as ASH. Sometimes materials do not get fragmented into ash but remain as larger pieces of semimolten magma with a solidified crust. Some of these materials are called VOLCANIC or LAVA BOMBS because of their size and the way in which they are thrown up out of the volcano to come crashing down on the surrounding countryside.

(Below) The material that comes from a volcano is always magma. However, whether the material flies high into the air in tiny pieces and creates ash, or whether it remains liquid as lava depends on the chemical composition and thus on the explosiveness of the magma. In this picture the eruption is throwing magma just high enough to form fist-sized lumps called cinders.

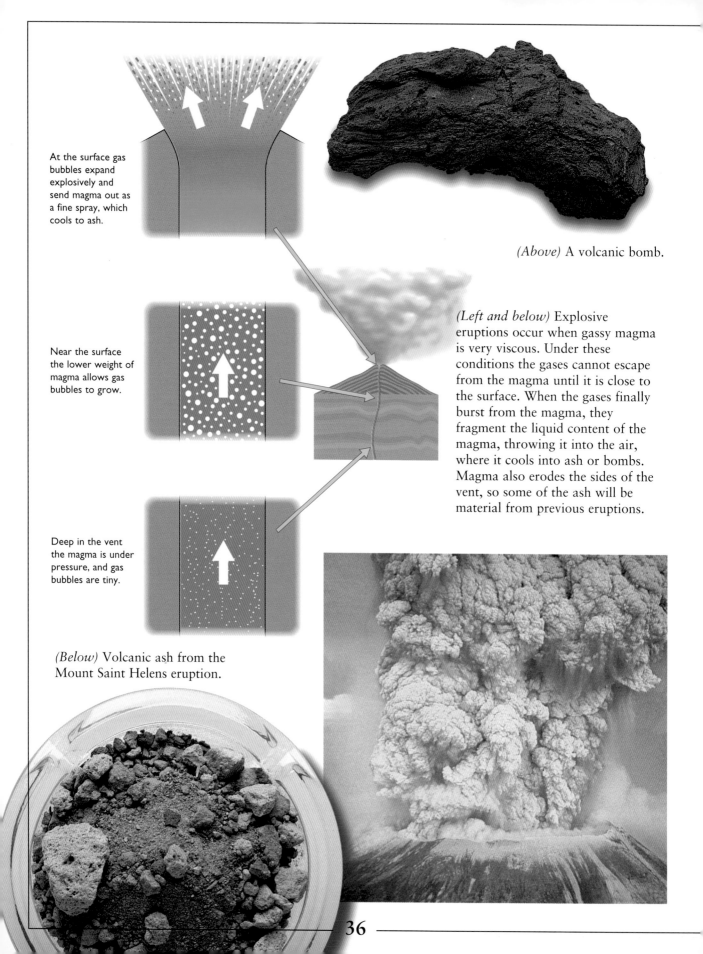

At the surface gas bubbles expand explosively and send magma out as a fine spray, which cools to ash.

(Above) A volcanic bomb.

Near the surface the lower weight of magma allows gas bubbles to grow.

(Left and below) Explosive eruptions occur when gassy magma is very viscous. Under these conditions the gases cannot escape from the magma until it is close to the surface. When the gases finally burst from the magma, they fragment the liquid content of the magma, throwing it into the air, where it cools into ash or bombs. Magma also erodes the sides of the vent, so some of the ash will be material from previous eruptions.

Deep in the vent the magma is under pressure, and gas bubbles are tiny.

(Below) Volcanic ash from the Mount Saint Helens eruption.

All solid (nonlava) materials ejected from a volcano are grouped together under the general term PYROCLASTICS, where *pyr* means fire, and *clastic* means broken. Sometimes gases and pieces of rock can act together as though they were liquid. In this, the most frightening form of eruption the material is called a PYROCLASTIC FLOW.

Volcanic rocks

No matter how sticky or runny the lava is when molten, eventually it will cool and become solid rock. As the lava cools, it often changes color. Basalt, for example, changes from red when liquid to black when solid.

Basalt is the most common form of lava. There are two contrasting kinds of basaltic lava. Lava that flows slowly and develops a thick skin and a rough, broken surface is called AA LAVA (a term borrowed from the Hawaiian language and pronounced ah-ah). Lava can also flow in thin sheets and develop a smooth surface. It is called PAHOEHOE LAVA (pronounced pa-howie-howie).

(Below) This diagram shows a side view through a lava flow. The aa lava is thick and develops a crust that breaks up as the lava below moves; the more runny pahoehoe lava moves forward in thin sheets that cool one by one.

The edge of an aa lava flow

Aa lava

Lava moving very slowly

Thick crust forms

Boulders of lava

Pahoehoe lava

Lava moving fairly quickly

Thin crust forms

In general, the stickier the lava, the rougher the surface it produces.

Although basalt is the most common form of lava, ANDESITE and RHYOLITE are also found in more acid volcanoes. All of these rocks contain a mixture of silica and iron-rich minerals such as the pyroxenes. When the proportion of silica to pyroxenes is low, the rock is very dark, develops very small crystals, and when molten, flows like water.

Basalt is the rock containing the least silica (usually less than half), and it flows out to make sheets of lava that may reach hundreds of kilometers from the fissure from which they were erupted.

Andesite is a gray, slightly more silica-rich rock (just over 50 percent silica) that also contains calcium feldspars. Andesite is associated with volcanoes that throw out volcanic bombs, but it is not the lava that comes from very explosive volcanoes. Andesite is less runny than basalt and forms tongues rather than sheets of lava that rarely travel much beyond the cone. Andesite is the most common material in stratovolcanoes (see page 40).

(Below) Solidified ash forms a rock called tuff.

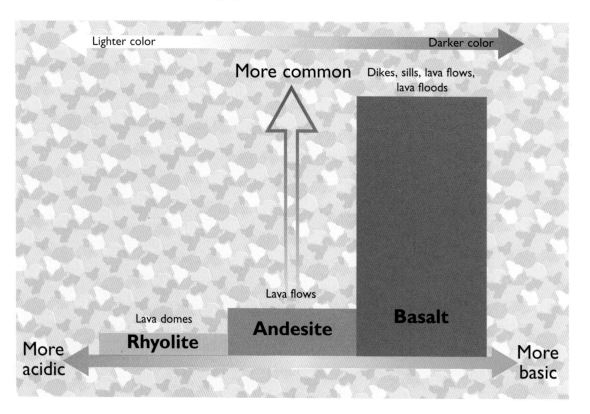

Lighter color → Darker color

More common

Dikes, sills, lava flows, lava floods

Lava flows

Lava domes

Rhyolite　　**Andesite**　　**Basalt**

More acidic ← → More basic

Dacite is a more coarsely grained rock, brown in color, not very common, and contains about two-thirds silica. The main feldspars are sodium feldspars. Dacite is associated with high viscosity lavas that produce steep-sided cones and that release large amounts of ash. It is linked with violent eruptions.

Rhyolite is a coarse-grained rock that forms the most sticky of all the lavas and makes only small lava domes. It is light gray or pink, contains about three-quarters silica, and is dominated by potassium feldspars. It is so sticky that it is usually associated with extremely violent eruptions.

Obsidian is a volcanic glass extremely rich in silica. It normally forms on the surfaces of very sticky lava.

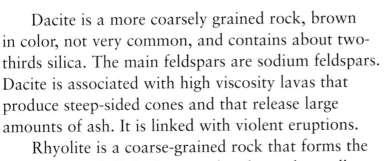
Rhyolite

Andesite

Basalt

(On this page) A selection of common volcanic rocks.

Shapes of volcanoes

The most perfectly shaped volcanoes are usually those with long histories of moderate explosions. Volcanoes with extremely violent histories are often less elegantly shaped because of the destruction that the eruption causes to the summit of the volcano.

Mount Fuji in Japan and Mount Mayon in the Philippines are often thought of as two of the most elegantly shaped volcanoes in the world. Their symmetrical concave-upward profiles are the result of thousands of moderately sized eruptions.

In most eruptions the early stages are the formation of ash, followed toward the end of the eruption by the ejection of tongues of lava. As a result, the volcano is made up of layers, or strata, of ash and lava overlying one another. That is the reason why such volcanoes are called STRATOVOLCANOES (made of many layers), or COMPOSITE VOLCANOES (composed of different types of rock), or CENTRAL VENT VOLCANOES (volcanoes whose material comes from a single central pipe or vent).

The less reliable and more violent volcanoes throw out a much wider variety of materials and so have a more complicated shape. They are often referred to as COMPLEX VOLCANOES. In this kind of volcano the longer the time between eruptions, the more likely it is that the next eruption will be very violent and explosive. Conversely, the sooner one eruption follows another, the less violent it will be. The reason for this is that the longer the volcano stays

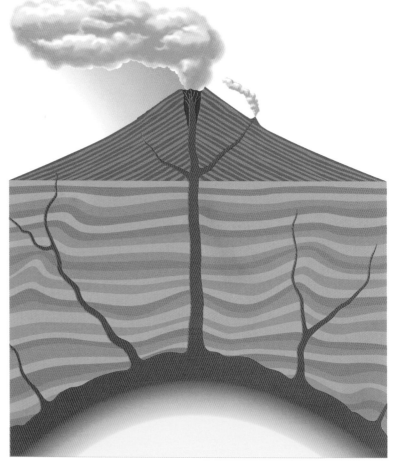

(*Below*) Volcanic cones may contain a variety of layers (strata) of material, often deposited one atop another. Thus many volcanoes of this kind are known as stratovolcanoes. If they have a single supply pipe, or vent, they will have a roughly symmetrical shape.

The magma chamber is located at a relatively shallow depth in the crust, and its presence weakens the crust, allowing magma to flow to the surface along many routes. Sometimes small vents open on the flanks of the volcano. When this happens, small cones build, and the main volcano loses its symmetrical shape.

inactive, the more the magma in the magma chamber below it can separate, and the more silica can rise to the top of the magma chamber. The higher the silica content of the magma, the more explosive the eruption.

When a volcano erupts under water, the shape of the volcano and its products are very different from those on land. In this case the pressure of the water tends to prevent any explosive activity, and most of the submarine volcanoes erupt low-viscosity magma in any case. The result is that the lava flows out quietly and is quickly cooled by the water, making a series of solidified bubbles, or pillow shapes. Submarine lava characteristically forms interlocking flows of pillow lava.

Because submarine volcanoes erupt quietly, they largely go unnoticed by people. Yet the eruptions from submarine volcanoes hugely exceed all the eruptions on land.

Stratovolcanoes

Stratovolcanoes are the most numerous of the world's volcanoes, making up 577 of the world's 1,300 active volcanoes. The upper parts of the cones of

(Above and below) Pillow lava on land. This rock was once part of the seabed and the lava erupted under a great depth of water. The pillows characteristically have an open center.

(Below) The stratovolcano of El Misti in the Peruvian Andes.

stratovolcanoes have steep sides with angles up to 35°. Most stratovolcanoes have a summit crater, and most contain andesite.

Shield volcanoes

SHIELD VOLCANOES are the second most numerous, with 107 of the world's total. The word shield is used because of their resemblance to the shape of a Hawaiian fighting shield

when resting on the ground. They are generally enormous volcanoes that have a dome shape (convex slopes, as opposed to the concave slopes of a stratovolcano). Most shield volcanoes contain only basalt. Investigations show that the number of layers making up a shield volcano is even larger than that for a stratovolcano, representing evidence of hundreds of thousands of eruptions. The largest of these volcanoes are very old, certainly over a million years. The summits of shield volcanoes are gentle, with slopes below 6°. They also have a very large

(Above) Mount Rainier, Washington, is a stratovolcano, made of both lava and ash layers. Notice that the sides of the cone have a concave slope. The volcano rises high above the general level of the Cascade Range in the northwestern part of the United States. Because the volcano has not erupted for some time, the summit and flanks have been eroded by glaciers.

(Left) The summit crater of Kilauea, on Hawaii. Notice the many layers making up the walls of the summit crater. Each one represents a separate major eruption.

summit crater, or series of craters following the line of a fissure. They are frequently filled with a pool of lava.

Seamounts

SEAMOUNTS are submarine volcanoes. They also comprise 107 of the world's known active volcanoes. Most seamounts occur along the submarine ridges that cross the great oceans. Most of the seamounts that send explosive materials into the air rise out of relatively shallow seas. There are certainly many more active seamounts than this, but they rise from the deep ocean floor and erupt entirely under water, and so their eruptions are thus far unknown. Loihi in Hawaii is an example of a deep seamount that has now risen fairly close to the surface of the sea. Its summit crater is one kilometer below the surface. It is the youngest volcano of the Hawaiian chain and will probably rise to be as great as Mauna Loa. Hidden seamounts are mostly detected through the earthquake swarms that mark each underwater eruption.

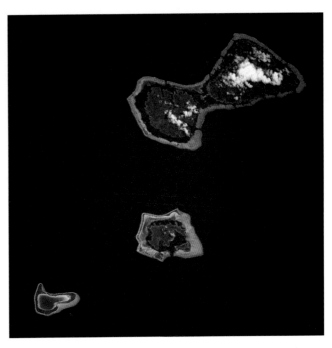

(Above) A space shuttle view of the seamount Bora Bora in the Pacific Ocean.

(Below) Looking from Tahiti across to Moorea in the Society Islands of French Polynesia. These reef-fringed mountainous islands are seamounts.

Cinder cones

CINDER CONES (also called SCORIA CONES) are small landforms made of loose fragments of material usually between pea and fist sized. The sides of the cones tend to stand at the angle of rest of coarse material, about 30°. These cinders are the remains of pyroclastic materials that were thrown out of a vent and cooled before landing. During the eruption the cinders would have been volcanic bombs, and the type of eruption would have been Hawaiian or Vulcanian (see pages 47 and 50). Most cinder cones are made from basaltic materials.

One of the most famous cinder cones in the world is Paricutin in Mexico, which formed a 410-meter-high cone between 1943 and 1952. The cone began to emerge quite unexpectedly one day in some cultivated farmland and continued to grow for nine years. Many cinder cones erupt in the same area to make a volcanic field. The Mountains of the Moon, Idaho, are a volcanic field.

The violence of an eruption

The separation of gases from the magma has very important effects on the nature of the volcanic eruption. The role of gases depends both on the amount of silica in the magma and on the amount of gas dissolved in it.

To understand this, a volcanic eruption can be compared with a bottle of soda. If the soda bottle is opened slowly, the carbon dioxide gas in the drink (which remains dissolved while the bottle contents are under pressure) can escape in a gradual way; when the cap is finally taken all the way off, the soda remains inside the bottle. If, on the other hand, the bottle is first shaken vigorously and then the cap removed quickly, the

(Below) The Mountains of the Moon National Monument in Idaho, in the northwestern United States, are a classic example of a region of cinder cones. The cones are built on pahoehoe and aa lava, showing that the material flowing to the surface was basalt.

gas comes out of solution very quickly, and the result is an explosive eruption of a mixture of liquid and gas.

In the case of quiescent lavas the basalt magma reaches the surface through long fissures, and the gases are able to escape easily. The lava is also runny, and this makes it easy for the gases to come out of solution. On the other hand, the magma such as that found at Mount Saint Helens (dacite lava) is so sticky that the gases cannot readily escape, and so the gases are not released until the lava is almost at the surface. As a result, they burst from the magma just as it is rising up the vent, and there is an explosive eruption of lava spray that cools in the air to form ash.

Chapter 4: Types of eruptions

Volcanic eruptions are classified according to their degree of eruptive violence. The violence associated with each type of volcano is a reflection of the materials ejected. However, it is important to notice that volcanoes that erupt with different degrees of violence may produce similar cones, so the shape of the cone is not a guide to the kind of eruption, except that the most gentle-sided volcanoes are connected with the least violent eruptions and also contain only low-viscosity basaltic lava. A stratovolcano commonly exhibits more than one kind of eruptive violence in its history, or even within the same eruption.

(Below) Fissure eruptions occur where tectonic plates pull apart and probably also where some hot spots reach the surface.

Icelandic or fissure type

The ICELANDIC TYPE of eruption rarely produces a volcanic cone, nor does lava flow out of a single pipe or vent. Instead, this type of eruption displays a flow of lava out of a fissure, so that a long split develops in the crust, and the lava simply pours out of it over a great length. These kinds of eruptions are most common at spreading boundaries, and most occur under water; but a few also occur on land, particularly in Iceland.

Iceland is entirely built of flood basalts. The largest eruption from the active Laki fissure was in 1783. Lava fountains occurred along its length, while lava poured out over the surrounding landscape, covering some 565 square kilometers and amounting to 12 cubic kilometers of new material.

Large fissure eruptions can pour out unimaginable floods of runny basalt, which can cover a landscape as fast as water. They

(Below) The Columbia River has cut through some of the flood basalts in the northwestern United States, producing a gorge.

have built up landscapes so immense that it is hard to realize that they are the result of a volcanic eruption. There have been no catastrophic eruptions from land fissures in recorded history. But fissure eruptions have in the past been immense. They also make up some of the world's best known landforms, including The Giant's Causeway in Northern Ireland and Fingal's Cave in Scotland. But far bigger than most people can imagine are the flood basalts that make up great areas of the Argentinian pampas and the majority of the Columbia-Snake basin in the northwestern United States. And biggest of all are the Deccan Traps, making up a landscape of basalt that covers about a sixth of the Indian subcontinent.

Hawaiian type

The Hawaiian chain of volcanic islands has given its name to the type of eruption that produces outpourings of basaltic lava from a single short fissure and therefore builds a broad, gently sloping cone. Many of these eruptions are associated with HOT SPOTS rather than with places where the crust is separating.

(Below) Basalt can readily flood out of fissures and cascade across the landscape.

The **HAWAIIAN TYPE** of eruption produces frequent short bursts of lava that form lava fountains. They act as the sources for rivers of lava that run across the landscape as individual tongues rather than as sheets. Many rivers of lava flow in lava tubes beneath a solidified crust of lava. Protected from losing heat by the overlying crust, they can flow for many kilometers. The volcanic cones produced by these eruptions are shield volcanoes. Mauna Loa, for example, is about 130 kilometers in diameter and reaches over 9,000 meters from the seafloor. The mountain has a volume of 40,000 cubic kilometers, making it the world's largest.

The Hawaiian type of eruption is more or less continually in action in a small way. The lava flows in Hawaii spread out radially from the summit.

(Left) Basaltic lava congeals on the surface into a black rock that looks like molten tar, while molten lava flows underneath in lava tubes, often for many kilometers.

(Above) Basalt lava flowing down the slopes of Mauna Loa, Hawaii.

(Below) This picture shows the summit of Kilauea near the top of the summit complex of Hawaii. The basaltic lava is ejected first as a fountain, creating a SPATTER CONE around the vent. The lava then flows away from the cone in the form of a river.

(Below) Hawaiian-type volcanoes typically produce a gently sloping cone.

Strombolian type

The **STROMBOLIAN TYPE** of eruption is sufficiently violent to produce a very high fountain of lava as well as sending rivers of lava down the sides of the cone. However, the eruptions are not regarded as being very threatening.

Stromboli, the volcano off the west coast of Italy after which the type is named, is known as the "lighthouse of the Mediterranean" because it erupts most of the time, and the glowing lumps of lava—called volcanic or lava bombs—that it thows out can be seen from offshore by night.

Strombolian eruptions, and all of the eruptions listed below, contribute to the steep-sided cones that are called stratovolcanoes.

Vulcanian type

The **VULCANIAN TYPE** of eruption is named after the Italian volcano Vulcano. It is more violent than the Strombolian type and is the first type to produce ash. Eruptions occur less often than in Strombolian; and when they occur, they are more violent, with clouds of gas and ash being thrown out in the form of a dark, swirling cloud over the volcano. The ash clouds, however, are not sent high into the atmosphere, and the ash does not travel very far.

Pelean type

The **PELEAN TYPE** is a violently explosive eruption in which clouds containing red-hot pieces of lava, together with gases, form a dense fluid that rolls down the side of the volcano, rather like a burning avalanche. The flow can exceed 100 km/hr. The original term used for such fast-moving, ground-hugging clouds was **NUÉE ARDENTE**, but the term pyroclastic flow is also widely used by geologists today. The first such eruption to be recorded was on Mount Pelée on

(Below) The Pelean type of eruption.

the Caribbean island of Martinique in 1902. One of the most recent was the eruption of Mount Saint Helens in the northwestern United States in 1980 (see page 55).

The majority of material erupted is gas and ash, not runny lava. A nuée ardente can travel for tens of kilometers. The density of the ash cloud is such that it can flatten most things in its path.

Plinian type

This is the most violent type of all the gas-cloud eruptions. It is named after a famous historical figure, Pliny the Elder, who was killed by the eruption of Mount Vesuvius in Italy in 79 AD. This eruption also destroyed the cities of Pompeii and Herculaneum. Present-day Naples lies at the foot of Mount Vesuvius.

In **PLINIAN TYPE** eruption the gases are more violently erupted than in the Pelean type, and they shoot up into the air in a giant gas and ash column. Ash from this kind of eruption can be carried by high-level winds and may remain in the air for months or years.

(Below) The Plinian type of eruption. Ash is the dominant form of material ejected from Plinian volcanoes, spreading out to blanket a large area of the downwind landscape.

(Above) Ash from the Mount Pinotubo, Philippines, Plinian eruption of 1991.

Mount Saint Helens: anatomy of a violent eruption

Mount Saint Helens was, until 1980, widely regarded as one of the most perfectly shaped conical volcanoes in the world. It rose from the Cascade Mountains in a beautifully symmetrical cone that was topped by a permanent snowfield and glaciers. Its summit was 2,950 meters high.

Geologists investigating Mount Saint Helens found evidence that the volcano had erupted at least 20 times in the previous 4,500 years, and that its last important period of eruption was in the mid-19th century. This was a sign that Mount Saint Helens was the kind of volcano that erupted only sporadically. That being the case, it was therefore likely to be the type that erupted violently.

As with all volcanoes, the first signs of an eruption were given by an increase in earthquake activity in the vicinity of the mountain. This showed that the pressure from the magma chamber below was causing the brittle rocks of the cone to begin to crack apart. Soon, enough lines of weakness would be created by the earthquakes for the top of the mountain to be dangerously unstable. The earthquakes were particularly numerous below the north flank of the volcano, indicating that it was the likely location of the eruption.

In fact, the first eruption was from the summit rather than from the flank and took place on March 27, just a week after the rapid increase in earthquakes. This small eruption continued in fits and starts thereafter. The material that came from the volcano was mainly gas and ash, and a small cloud of material was sent up over the cone. At

(Above) Mount Saint Helens is in the Cascade Mountains in the northwest United States.

(Below) Mount Saint Helens (foreground) and other volcanic peaks in the Cascade Mountain chain before the eruption of 1980.

this stage it was a Vulcanian type of eruption; but because the size of the eruption was small, the ash fell down to the ground close to the summit, and the main effect was simply to blacken the snowfields.

The cause of these eruptions was probably water from the icefields seeping down to the hot rock below and then flashing into steam. The effect of this explosive change from water to steam was to throw out material from the vent. But this was all old material, and no new magma was involved at this stage.

As the days went by, a large bulge, about two kilometers across, appeared to be forming on the northern flank of the mountain, separate from the central vent. It showed that the material blocking the central vent was stuck fast, and that the magma was seeking a way out through the northern flank.

Although geologists could not know this at the time, the earthquakes had weakened the whole

(Below) The destruction was to the north of the volcano because the new eruption was on the northern flank of the cone. Nonetheless, the whole top of the cone was blown away as well as a large part of the northern flank. The destructive power of the volcano is seen in the satellite image by the contrast between the forested (green) southern flank and the large area of barren land (brown) to the north of the volcano. The feature in the center of the cone is a lava dome that slowly built up in the years after the explosive eruption. It is made of a very acid form of lava.

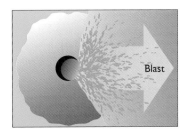

Spirit lake

53

north flank more than any other part of the cone. The rock was simply held together by friction. As the bulge grew, the rock was being pushed apart and replaced by liquid magma, so that the cone was liable to fail at any moment.

The actual moment of failure was recorded in the form of a much larger earthquake on

(Below) The pyroclastic flow gets under way in this spectacular eruption from Mount Saint Helens.

(Above) Eventually the volcano sent material high into the sky. This is a Plinian type of eruption.

May 18 just after 8 A.M. At this time the entire north flank of the volcano dropped down in a massive landslide of two cubic kilometers of rock. It was as though the cap had been taken off a shaken soda bottle (see page 36). Now, with nothing to hold it in place, the gas-rich magma was able to escape in a gigantic explosion.

In fact the landslide was shattered into tiny fragments by the explosion, so that a mixture of rock fragment and gas was formed. This part of the eruption was now a Pelean type, and the gas and pyroclastic materials were able to move at speeds of up to 250 kilometers an hour. Part of this pyroclastic flow went into nearby Spirit Lake, throwing out the water entirely, while the rest went off down a valley, traveling some 28 kilometers before coming to rest.

The blast demolished the huge forest that lay to the north of the volcano, leveling everything over an area of 550 square kilometers. Close to the volcano the force of the blast was so great that the trees were simply blown away. Beyond this was a zone where everything was knocked down. In total some ten million trees were felled by the explosion.

The Pelean eruption was, however, only the first part of the total eruption. It had taken off the top of the mountain and created a huge crater in the north flank from which a new eruption then occurred.

As soon as a new vent was established, the gases and pyroclastic materials were able to blow upward, creating a Plinian type of eruption. This ash rose up into the stratosphere and was carried by the winds eastward. Ash was then deposited over a wide area that included the town of Yakima.

By the time the eruption was over, the volcano had been reduced in height from 2,900 meters to 2,500 meters, and some 2.7 cubic kilometers of material had been moved, of which about half a cubic kilometer was new magma. The rest was old cone material blown away in the first blasts.

In the following days the Plinian eruption subsided, but instead a very sticky lava began to flow into the new crater, building up a lava dome on the crater floor. This dome, which caps the new vent, remains the final feature of the eruption. It continued to grow until 1986.

But devastating as these eruptions were, the violence of the Mount Saint Helens eruption spread much further than the area blasted away or buried by ash. This was because the glaciers and snowfields on the mountain had been turned into steam, and they were now coming back to the ground in the form of torrential rainfall. The rain swept much of the ash into the nearby rivers, causing widespread flooding and at the same time silting them up.

Mount Saint Helens was not a large eruption. There have been some far more devastating eruptions, such as when the Indonesian island of Krakatoa blew apart in 1883, expelling 100 cubic kilometers of material and creating a sound that could be heard across the world.

(Below) The desolate, ash-covered landscape left by the Mount Saint Helens eruption. This view looks southwest across Spirit Lake toward the crater.

Kilauea: quiet, constant eruption

Shield volcanoes erupt most of the time without explosive violence. There is therefore never a single catastrophic event for people to witness.

In 1984, for example, the eruption that began at Kilauea, Hawaii, started after a period in which small earthquakes had become more frequent. When the eruption came, a fissure appeared in the large summit crater (called a caldera, see page 58), and fountains of lava 50 meters high spurted out from all along the crack. Within a few hours the lava had covered the bottom of the caldera, which was about three by five kilometers in size.

The next day a long fissure opened up on the northeast side of the volcano's flank, then another crack opened. As each new crack appeared, lava spurted up as a curtain of fiery orange liquid.

In this case the opening of the lower fissures made the upper fissures stop erupting as though they were tapping the magma chamber at lower and lower levels.

The result of these fissures was to produce lava flows reaching some 27 kilometers long and yielding a total of about 220,000 cubic meters of lava that covered an area of nearly 50 square kilometers.

(Below) The location of Kilauea on Hawaii.

(Below) A lava fountain accompanied by ash, Kilauea.

(Below) Lava flowing from Kilauea.

Calderas and crater lakes

Some volcanoes do not rise gracefully to a tall cone with a small crater at the summit. Instead, the top of the cone is missing, and the "crater" is immensely wide. A very large crater is called a CALDERA. Calderas often contain a large lake, named a CRATER LAKE.

The caldera is very large because the top of the cone has collapsed back in on itself. So while a crater is simply the top of a vent, a caldera represents the collapsed top of a volcano.

To make a caldera, the magma chamber feeding the volcano has to be quite close to the surface. Then the pressure from the chamber can make the rocks above it crack and weaken. During an eruption some of the contents of the magma chamber are used up, so the

(Below) A satellite near-vertical image of Pico de Teide, Tenerife, Canary Islands. The half-oval shape in the center of the picture is a summit caldera that is about 16 kilometers across at its widest. The half of the caldera facing top right has been blown away.

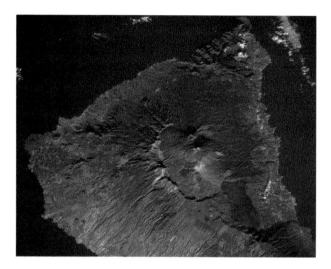

(Below and right) This is Crater Lake in Oregon, probably the world's most famous crater lake. The volcano used to be 4,500 meters high, but 6,500 years ago the summit collapsed. The rim of Crater Lake is now about 2,000 meters above sea level, so more than half of the mountain sank into the ground. A new cone has been building up since this time. It forms Wizard Island in the middle of this picture.

chamber is partly empty. Above it lies weak rock, and on top of this is the immense weight of the volcanic mountain.

The weight of the volcano can be enough to collapse the whole center of the mountain into the chamber below.

After the eruption is over, all that is left of the top of the mountain is a huge pit that, if the rim is intact, may fill with water to make a crater lake.

One of the most famous crater lakes is in Oregon. The volcano that used to exist there has been called Mount Mazama by scientists. The summit lake—called Crater Lake—is 11 kilometers across, and the water is over 600 meters deep.

Just because the center of the volcano collapses, it does not mean that the volcano becomes extinct. Far from it. In the case of Crater Lake a new cone—Wizard Island—is already forming.

(Below) Stages in the formation of a crater lake.

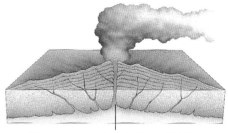

The volcano erupts violently, so that the rocks are weakened. At the same time, the lava pours out of the vent so fast that the magma chamber is left partly empty.

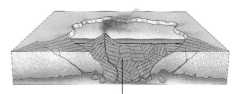

The weight of the volcano's cone makes it collapse in on the magma chamber, producing a vast pit that can then, over the following centuries, fill with water.

Glossary

aa lava: a type of lava with a broken, bouldery surface.

abrasion: the rubbing away (erosion) of a rock by the physical scraping of particles carried by water, wind, or ice.

acidic rock: a type of igneous rock that consists predominantly of light-colored minerals and more than two-thirds silica (e.g., granite).

active volcano: a volcano that has observable signs of activity, for example, periodic plumes of steam.

adit: a horizontal tunnel drilled into rock.

aftershock: an earthquake that follows the main shock. Major earthquakes are followed by a number of aftershocks that decrease in frequency with time.

agglomerate: a rock made from the compacted particles thrown out by a volcano (e.g., tuff).

alkaline rock: a type of igneous rock containing less than half silica and normally dominated by dark-colored minerals (e.g., gabbro).

amygdule: a vesicle in a volcanic rock filled with such secondary minerals such as calcite, quartz, or zeolite.

andesite: an igneous volcanic rock. Slightly more acidic than basalt.

anticline: an arching fold of rock layers in which the rocks slope down from the crest. *See also* syncline.

Appalachian Mountain (Orogenic) Belt: an old mountain range that extends for more than 3,000 km along the eastern margin of North America from Alabama in the southern United States to Newfoundland, Canada, in the north. There were three Appalachian orogenies: Taconic (about 460 million years ago) in the Ordovician; Acadian (390 to 370 million years ago) in the Devonian; and Alleghenian (300 to 250 million years ago) in the Late Carboniferous to Permian. These mountain belts can be traced as the Caledonian and Hercynian orogenic belts in Europe.

Archean Eon: *see* eon.

arenaceous: a rock composed largely of sand grains.

argillaceous: a rock composed largely of clay.

arkose: a coarse sandstone formed by the disintegration of a granite.

ash, volcanic: fine powdery material thrown out of a volcano.

asthenosphere: the weak part of the upper mantle below the lithosphere, in which slow convection is thought to take place.

augite: a dark green-colored silicate mineral containing calcium, sodium, iron, aluminum, and magnesium.

axis of symmetry: a line or plane around which one part of a crystal is a mirror image of another part.

basalt: basic fine-grained igneous volcanic rock; lava often contains vesicles.

basic rock: an igneous rock (e.g., gabbro) with silica content less than two-thirds and containing a high percentage of dark-colored minerals.

basin: a large, circular, or oval sunken region on the earth's surface created by downward folding. A river basin, or watershed, is the area drained by a river and its tributaries.

batholith: a very large body of plutonic rock that was intruded deep into the earth's crust and is now exposed by erosion.

bauxite: a surface material that contains a high percentage of aluminum silicate. The principal ore of aluminum.

bed: a layer of sediment. It may involve many phases of deposition, each marked by a bedding plane.

bedding plane: an ancient surface on which sediment built up. Sedimentary rocks often split along bedding planes.

biotite: a black-colored form of mica.

body wave: a seismic wave that can travel through the interior of the earth. P waves and S waves are body waves.

boss: an upward extension of a batholith. A boss may once have been a magma chamber.

botryoidal: the shape of a mineral that resembles a bunch of grapes, e.g., hematite whose crystals are often arranged in massive clumps, giving a surface covered with spherical bulges.

butte: a small mesa.

calcareous: composed mainly of calcium carbonate.

calcite: a mineral composed of calcium carbonate.

caldera: the collapsed cone of a volcano. It sometimes contains a crater lake.

Caledonian Mountain-Building Period, Caledonian Orogeny: a major mountain-building period in the Lower Paleozoic Era that reached its climax at the end of the Silurian Period (430 to 395 million years ago). An early phase affected only North America and made part of the Appalachian Mountain Belt.

Cambrian, Cambrian Period: the first period of geological time in the Paleozoic Era, beginning 570 million years ago and ending 500 million years ago.

carbonate minerals: minerals formed with carbonate ions (e.g., calcite).

Carboniferous, Carboniferous Period: a period of geological time between about 345 and 280 million years ago. It is often divided into the Early Carboniferous Epoch (345 to 320 million years ago) and the Late Carboniferous Epoch (320 to 280 million years ago). The Late Carboniferous is characterized by large coal-forming swamps. In North America the Carboniferous is usually divided into the Mississippian (= Lower Carboniferous) and Pennsylvanian (= Upper Carboniferous) periods.

cast, fossil: the natural filling of a mold by sediment or minerals that were left when a fossil dissolved after being enclosed by rock.

Cenozoic, Cenozoic Era: the most recent era of geological time, beginning 65 million years ago and continuing to the present.

central vent volcano: *see* stratovolcano

chemical compound: a substance made from the chemical combination of two or more elements.

chemical rock: a rock produced by chemical precipitation (e.g., halite).

chemical weathering: the decay of a rock through the chemical action of water containing dissolved acidic gases.

cinder cone: a volcanic cone made entirely of cinders. Cinder cones have very steep sides.

class: the level of biological classification below a phylum.

clast: an individual grain of a rock.

clastic rock: a sedimentary rock that is made up of fragments of preexisting rocks, carried by gravity, water, or wind (e.g., conglomerate, sandstone).

cleavage: the tendency of some minerals to break along one or more smooth surfaces.

coal: the carbon-rich, solid mineral derived from fossilized plant remains. Found in sedimentary rocks. Types of coal include bituminous, brown, lignite, and anthracite. A fossil fuel.

complex volcano: a volcano that has had an eruptive history that produces two or more vents.

composite volcano: *see* stratovolcano.

concordant coast: a coast where the geological structure is parallel to the coastline. *See also* discordant coastline.

conduction (of heat): the transfer of heat between touching objects.

conglomerate: a coarse-grained sedimentary rock with grains larger than 2 mm.

contact metamorphism: metamorphism that occurs due to direct contact with a molten magma. *See also* regional metamorphism.

continental drift: the theory suggested by Alfred Wegener that earth's continents were originally one land mass that split up to form the arrangement of continents we see today.

continental shelf: the ocean floor from the coastal shore of continents to the continental slope.

continental shield: the ancient and stable core of a tectonic plate. Also called a shield.

convection: the slow overturning of a liquid or gas that is heated from below.

cordillera: a long mountain belt consisting of many mountain ranges.

core: the innermost part of the earth. The earth's core is very dense, rich in iron, partly molten, and the source of the earth's magnetic field. The inner core is solid and has a radius of about 1,300 kilometers. The outer core is fluid and is about 2,100 kilometers thick. S waves cannot travel through the outer core.

cracking: the breaking up of a hydrocarbon compound into simpler constituents by means of heat.

crater lake: a lake found inside a caldera.

craton: *see* shield.

Cretaceous, Cretaceous Period: the third period of the Mesozoic Era. It lasted between about 135 and 65 million years ago. It was a time of chalk formation and when many dinosaurs lived.

cross-bedding: a pattern of deposits in a sedimentary rock in which many thin layers lie at an angle to the bedding planes, showing that the sediment was deposited by a moving fluid. Wind-deposited cross-beds are often bigger than water-deposited beds.

crust: the outermost layer of the earth, typically 5 km under the oceans and 50 to 100 km thick under continents. It makes up less than 1 percent of the earth's volume.

crustal plate: *see* tectonic plate.

crystal: a mineral that has a regular geometric shape and is bounded by smooth, flat faces.

crystal system: a group of crystals with the same arrangement of axes.

crystalline: a mineral that has solidified but been unable to produce well-formed crystals. Quartz and halite are commonly found as crystalline masses.

crystallization: the formation of crystals.

cubic: a crystal system in which crystals have 3 axes all at right angles to one another and of equal length.

cuesta: a ridge in the landscape formed by a resistant band of dipping rock. A cuesta has a steep scarp slope and a more gentle dip slope.

current bedding: a pattern of deposits in a sedimentary rock in which many thin layers lie at an angle to the bedding planes, showing that the sediment was deposited by a current of water.

cyclothem: a repeating sequence of rocks found in coal strata.

delta: a triangle of deposition produced where a river enters a sea or lake.

deposit, deposition: the process of laying down material that has been transported in suspension or solution by water, ice, or wind. A deposit is the material laid down by deposition (e.g., salt deposits).

destructive plate boundary: a line where plates collide and one plate is subducted into the mantle.

Devonian, Devonian Period: the fourth period of geological time in the Paleozoic Era from 395 to 345 million years ago.

dike: a wall-like sheet of igneous rock that cuts across the layers of the surrounding rocks.

dike swarm: a collection of hundreds or thousands of parallel dikes.

diorite: an igneous plutonic rock between gabbro and granite; the plutonic equivalent of andesite.

dip: the angle that a bedding plane or fault makes with the horizontal.

dip slope: the more gently sloping part of a cuesta whose surface often parallels the dip of the strata.

discontinuity: a gap in deposition, perhaps caused by the area being lifted above the sea so that erosion, rather than deposition, occurred for a time.

discordant coast: a coast where the rock structure is at an angle to the line of the coast. *See also* concordant coastline.

displacement: the distance that one piece of rock is pushed relative to another.

dissolve: to break down a substance into a solution without causing a reaction.

distillation: the boiling off of volatile materials, leaving a residue.

dolomite: a mineral composed of calcium magnesium carbonate.

dome: a circular uplifted region of rocks taking the shape of a dome and found in some areas of folded rocks. Rising plugs of salt will also dome up the rocks above them. They sometimes make oil traps.

dormant volcano: a volcano that shows no signs of activity but that has been active in the recent past.

drift: a tunnel drilled in rock and designed to provide a sloping route for carrying out ore or coal by means of a conveyor belt.

earthquake: shaking of the earth's surface caused by a sudden movement of rock within the earth.

element: a fundamental chemical building block. A substance that cannot be separated into simpler substances by any chemical means. Oxygen and sulfur are examples of elements.

eon: the largest division of geological time. An eon is subdivided into eras. Precambrian time is divided into the Archean (earlier than 2.5 billion years ago) and Proterozoic eons (more recent than 2.5 billion years ago). The Phanerozoic Eon includes the Cambrian Period to the present.

epicenter: the point on the earth's surface directly above the focus (hypocenter) of an earthquake.

epoch: a subdivision of a geological period in the geological time scale (e.g., Pleistocene Epoch).

era: a subdivision of a geological eon in the geological time scale (e.g., Cenozoic Era). An era is subdivided into periods.

erode, erosion: the twin processes of breaking down a rock (called weathering) and then removing the debris (called transporting).

escarpment: the crest of a ridge made of dipping rocks.

essential mineral: the dominant mineral constituents of a rock used to classify it.

evaporite: a mineral or rock formed as the result of evaporation of salt-laden water, such as a lagoon or salt lake.

exoskeleton: another word for shell. Applies to invertebrates.

extinct volcano: a volcano that has shown no signs of activity in historic times.

extrusive rock, extrusion: an igneous volcanic rock that has solidified on the surface of the earth.

facet: the cleaved face of a mineral. Used in describing jewelry.

facies: physical, chemical, or biological variations in a sedimentary bed of the same geological age (e.g., sandy facies, limestone facies).

family: a part of the classification of living things above a genus.

fault: a deep fracture or zone of fractures in rocks along which there has been displacement of one side relative to the other. It represents a weak point in the crust and upper mantle.

fault scarp: a long, straight, steep slope in the landscape that has been produced by faulting.

feldspar: the most common silicate mineral. It consists of two forms: plagioclase and orthoclase.

ferromagnesian mineral: dark-colored minerals such as augite and hornblende that contain relatively high proportions of iron and magnesium and low proportions of silica.

fissure: a substantial crack in a rock.

fjord: a glaciated valley in a mountainous area coastal area that has been partly flooded by the sea.

focal depth: the depth of an earthquake focus below the surface.

focus: the origin of an earthquake, directly below the epicenter.

fold: arched or curved rock strata.

fold axis: line following the highest arching in an anticline or the lowest arching in a syncline.

fold belt: a part of a mountain system containing folded sedimentary rocks.

foliation: a texture of a rock (usually schist) that resembles the pages in a book.

formation: a word used to describe a collection of related rock layers, or beds. A number of related beds make a member; a collection of related members makes up a formation. Formations are often given location names, e.g., Toroweap Formation, whose members are a collection of dominantly limestone beds.

fossil: any evidence of past life, including remains, traces, and imprints.

fossil fuel: any fuel that was formed in the geological past from the remains of living organisms. The main fossil fuels are coal and petroleum (oil and natural gas).

fraction: one of the components of crude oil that can be separated from others by heating and then cooling the vapor.

fracture: a substantial break across a rock.

fracture zone: a region in which fractures are common. Fracture zones are particularly common in folded rock and near faults.

frost shattering: the process of breaking pieces of rock through the action of freezing and melting of rainwater

gabbro: alkaline igneous plutonic rock, typically showing dark-colored crystals; plutonic equivalent of basalt.

gallery: a horizontal access tunnel in a mine.

gangue: the unwanted mineral matter found in association with a metal.

gem: a mineral, usually in crystal form, that is regarded as having particular beauty and value.

genus: (*pl.* genera) the biological classification for a group of closely related species.

geode: a hollow lump of rock (nodule) that often contains crystals.

geological column: a columnar diagram showing the divisions of geological time (eons, eras, periods, and epochs).

geological eon: *see* eon.

geological epoch: *see* epoch.

geological era: *see* era.

geological period: a subdivision of a geological era (e.g., Carboniferous Period). A period is subdivided into epochs.

geological system: a term for an accumulation of strata that occurs during a geological period (e.g., the Ordovician System is the rocks deposited during the Ordovician Period). Systems are divided into series.

geological time: the history of the earth revealed by its rocks.

geological time scale: the division of geological time into eons, era, periods, and epochs.

geosyncline: a large, slowly subsiding region marginal to a continent where huge amounts of sediment accumulate. The rocks in a geosyncline eventually are lifted to form mountain belts.

gneiss: a metamorphic rock showing large grains.

graben: a fallen block of the earth's crust forming a long trough separated on all sides by faults. Associated with rift valleys.

grain: a particle of a rock or mineral.

granite: an acidic, igneous, plutonic rock containing free quartz, typically light in color; plutonic equivalent of rhyolite.

grit: grains larger than sand but smaller than stones.

groundmass: *see* matrix.

group: a word used to describe a collection of related rock layers, or beds. A number of related beds make a member; a collection of related members makes up a formation; a collection of related formations makes a group.

gypsum: a mineral made of calcium sulfate.

halide minerals: a group of minerals (e.g., halite) that contain a halogen element (elements similar to chlorine) bonded with another element. Many are evaporite minerals.

halite: a mineral made of sodium chloride.

Hawaiian-type eruption: a name for a volcanic eruption that mainly consists of lava fountains.

hexagonal: a crystal system in which crystals have 3 axes all at 120 degrees to one another and of equal length.

hogback: a cuesta where the scarp and dip slopes are about the same angle.

hornblende: a dark-green silicate mineral of the amphibole group containing sodium, potassium, calcium, magnesium, iron, and aluminum.

horst: a raised block of the earth's crust separated on all sides by faults. Associated with rift valleys.

hot spot: a place where a fixed mantle magma plume reaches the surface.

hydraulic action: the erosive action of water pressure on rocks.

hydrothermal: a change brought about in a rock or mineral due to the action of superheated mineral-rich fluids, usually water.

hypocenter: the calculated location of the focus of an earthquake.

ice wedging: *see* frost shattering

Icelandic-type eruption: a name given to a fissure type of eruption.

igneous rock: rock formed by the solidification of magma. Igneous rocks include volcanic and plutonic rocks.

impermeable: a rock that will not allow a liquid to pass through it.

imprint: a cast left by a former life form.

impurities: small amounts of elements or compounds in an otherwise homogeneous mineral.

index fossil: a fossil used as a marker for a particular part of geological time.

intrusive rock, intrusion: rocks that have formed from cooling magma below the surface. When inserted amoung other rocks, intruded rocks are called an intrusion.

invertebrate: an animal with an external skeleton.

ion: a charged particle.

island arc: a pattern of volcanic islands that follow the shape of an arc when seen from above.

isostacy: the principle that a body can float in a more dense fluid. The same as buoyancy, but used for continents.

joint: a significant crack between blocks of rock, normally used in the context of patterns of cracks.

Jurassic, Jurassic Period: the second geological period in the Mesozoic Era, lasting between 190 and 135 million years ago.

kingdom: the broadest division in the biological classification of living things.

laccolith: a lens-shaped body of intrusive igneous rock with a dome-shaped upper surface and a flat bottom surface.

landform: a recognizable shape of part of the landscape, for example, a cuesta.

landslide: the rapid movement of a slab of soil down a steep hillslope.

lateral fault: *see* thrust fault.

laterite: a surface deposit containing a high proportion of iron.

lava: molten rock material extruded onto the surface of the earth.

lava bomb: *see* volcanic bomb.

law of superposition: the principle that younger rock is deposited on older.

limestone: a carbonate sedimentary rock composed of more than half calcium carbonate.

lithosphere: that part of the crust and upper mantle that is brittle and makes up the tectonic plates.

lode: a mining term for a rock containing many rich ore-bearing minerals. Similar to vein.

Love wave, L wave: a major type of surface earthquake wave that shakes the ground surface at right angles to the direction the wave is traveling in. It is named after A.E.H. Love, the English mathematician who discovered it.

luster: the way in which a mineral reflects light. Used as a test when identifying minerals.

magma: the molten material that comes from the mantle and that cools to form igneous rocks.

magma chamber: a large cavity melted in the earth's crust and filled with magma. Many magma chambers are plumes of magma that have melted their way from the mantle to the upper part of the crust. When a magma chamber is no longer supplied with molten magma, the magma solidifies to form a granite batholith.

mantle: the layer of the earth between the crust and the core. It is approximately 2,900 kilometers thick and is the largest of the earth's major layers.

marginal accretion: the growth of mountain belts on the edges of a shield.

mass extinction: a time when the majority of species on the planet were killed off.

matrix: the rock or sediment in which a fossil is embedded; the fine-grained rock in which larger particles are embedded, for example, in a conglomerate.

mechanical weathering: the disintegration of a rock by frost shattering/ice wedging.

mesa: a large detached piece of a tableland.

Mesozoic, Mesozoic Era: the geological era between the Paleozoic and the Cenozoic eras. It lasted between 225 and 65 million years ago.

metamorphic aureole: the region of contact metamorphic rock that surrounds a batholith.

metamorphic rock: any rock (e.g., schist, gneiss) that was formed from a preexisting rock through heat and pressure.

meteorite: a substantial chunk of rock in space.

micas: a group of soft, sheetlike silicate minerals (e.g., biotite, muscovite).

midocean ridge: a long mountain chain on the ocean floor where basalt periodically erupts, forming new oceanic crust.

mineral: a naturally occurring inorganic substance of definite chemical composition (e.g., calcite, calcium carbonate).
   More generally, any resource extracted from the ground by mining (includes, metal ores, coal, oil, gas, rocks, etc.).

mineral environment: the place where a mineral or a group of associated minerals forms. Mineral environments include igneous, sedimentary, and metamorphic rocks.

mineralization: the formation of minerals within a rock.

Modified Mercalli Scale: a scale for measuring the impact of an earthquake. It is composed of 12 increasing levels of intensity that range from imperceptible, designated by Roman numeral I, to catastrophic destruction, designated by XII.

Mohorovicic discontinuity: the boundary surface that separates the earth's crust from the underlying mantle. Named for Andrija Mohorovicic, a Croatian seismologist.

Mohs' Scale of Hardness: a relative scale developed to put minerals into an order. The hardest is 10 (diamond), and the softest is 1 (talc).

mold: an impression in a rock of the outside of an organism.

monoclinic: a crystal system in which crystals have 2 axes all at right angles to one another, and each axis is of unequal length.

mountain belt: a region where there are many ranges of mountains. The term is often applied to a wide belt of mountains produced during mountain building.

mountain building: the creation of mountains as a result of the collision of tectonic plates. Long belts or chains of mountains can form along the edge of a continent during this process. Mountain building is also called orogeny.

mountain-building period: a period during which a geosyncline is compressed into fold mountains by the collision of two tectonic plates. Also known as orogenesis.

mudstone: a fine-grained, massive rock formed by the compaction of mud.

nappe: a piece of a fold that has become detached from its roots during intensive mountain building.

native metal: a metal that occurs uncombined with any other element.

natural gas: *see* petroleum.

normal fault: a fault in which one block has slipped down the face of another. It is the most common kind of fault and results from tension.

nuée ardente: another word for pyroclastic flow.

ocean trench: a deep, steep-sided trough in the ocean floor caused by the subduction of oceanic crust beneath either other oceanic crust or continental crust.

olivine: the name of a group of magnesium iron silicate minerals that have an olive color.

order: a level of biological classification between class and family.

Ordovician, Ordovician Period: the second period of geological time within the Paleozoic Era. It lasted from 500 to 430 million years ago.

ore: a rock containing enough useful metal or fuel to be worth mining.

ore mineral: a mineral that occurs in sufficient quantity to be mined for its metal. The compound must also be easy to process.

organic rocks: rocks formed by living things, for example, coal.

orthoclase: the form of feldspar that is often pink in color and that contains potassium as important ions.

orogenic belt: a mountain belt.

orogeny: a period of mountain building. Orogenesis is the process of mountain building and the creation of orogenic belts.

orthorhombic: a crystal system in which crystals have 3 axes all at right angles to one another but of unequal length.

outcrop: the exposure of a rock at the surface of the earth.

overburden: the unwanted layer(s) of rock above an ore or coal body.

oxide minerals: a group of minerals in which oxygen is a major constituent. A compound in which oxygen is bonded to another element or group.

Pacific Ring of Fire: the ring of volcanoes and volcanic activity that circles the Pacific Ocean. Created by the collision of the Pacific Plate with its neighboring plates.

pahoehoe lava: the name for a form of lava that has a smooth surface.

Paleozoic, Paleozoic Era: a major interval of geological time. The Paleozoic is the oldest era in which fossil life is commonly found. It lasted from 570 to 225 million years ago.

paleomagnetism: the natural magnetic traces that reveal the intensity and direction of the earth's magnetic field in the geological past.

pegmatite: an igneous rock (e.g., a dike) of extremely coarse crystals.

Pelean-type eruption: a violent explosion dominated by pyroclastic flows.

period: *see* geological period.

permeable rock: a rock that will allow a fluid to pass through it.

Permian, Permian Period: the last period of the Paleozoic Era, lasting from 280 to 225 million years ago.

petrified: when the tissues of a dead plant or animal have been replaced by minerals, such as silica, they are said to be petrified (e.g., petrified wood).

petrified forest: a large number of fossil trees. Most petrified trees are replaced by silica.

petroleum: the carbon-rich, and mostly liquid, mixture produced by the burial and partial alteration of animal and plant remains. Petroleum is found in many sedimentary rocks. The liquid part of petroleum is called oil. The gaseous part of petroleum is known as natural gas. Petroleum is an important fossil fuel.

petroleum field: a region from which petroleum can be recovered.

Phanerozoic Eon: the most recent eon, beginning at the Cambrian Period, some 570 million years ago, and extending up to the present.

phenocryst: an especially large crystal (in a porphyritic rock) embedded in smaller mineral grains.

phylum: (*pl.* phyla) biological classification for one of the major divisions of animal life and second in complexity to kingdom. The plant kingdom is not divided into phyla but into divisions.

placer deposit: a sediment containing heavy metal grains (e.g., gold) that have weathered out of the bedrock and concentrated on a stream bed or along a coast.

plagioclase: the form of feldspar that is often white or gray and that contains sodium and calcium as important ions.

planetismals: small embryo planets.

plate: *see* plate tectonics, tectonic plate.

plateau: an extensive area of raised flat land. The clifflike edges of a plateau may, when eroded, leave isolated features such as mesas and buttes. *See also* tableland.

plate tectonics: the theory that the earth's crust and upper mantle (the lithosphere) are broken into a number of more or less rigid, but constantly moving, slabs or plates.

Plinian-type eruption: an explosive eruption that sends a column of ash high into the air.

plug: *see* volcanic plug

plunging fold: a fold whose axis dips, or plunges, into the ground.

plutonic rock: an igneous rock that has solidified at great depth and contains large crystals due to the slowness of cooling (e.g., granite, gabbro).

porphyry, porphyritic rock: an igneous rock in which larger crystals (phenocrysts) are enclosed in a fine-grained matrix.

Precambrian, Precambrian time: the whole of earth history before the Cambrian Period. Also called Precambrian Era and Precambrian Eon.

precipitate: a substance that has settled out of a liquid as a result of a chemical reaction between two chemicals in the liquid.

Primary Era: an older name for the Paleozoic Era.

prismatic: a word used to describe a mineral that has formed with one axis very much longer than the others.

Proterozoic Eon: *see* eon.

P wave, primary wave, primary seismic wave: P waves are the fastest body waves. The waves carry energy in the same line as the direction of the wave. P waves can travel through all layers of the earth and are generally felt as a thump. *See also* S wave.

pyrite: iron sulfide. It is common in sedimentary rocks that were poor in oxygen and sometimes forms fossil casts.

pyroclastic flow: solid material ejected from a volcano, combined with searingly hot gases, which together behave as a high-density fluid. Pyroclastic flows can do immense damage, as was the case with Mount Saint Helens.

pyroclastic material: any solid material ejected from a volcano.

Quaternary, Quaternary Period: the second period in the Cenozoic Era, beginning about 1.6 million years ago and continuing to the present day.

radiation: the transfer of energy between objects that are not in contact.

radioactive dating: the dating of a material by the use of its radioactive elements. The rate of decay of any element changes in a predictable way, allowing a precise date to be given since the material was formed.

rank: a name used to describe the grade of coal in terms of its possible heat output. The higher the rank, the more the heat output.

Rayleigh wave: a type of surface wave having an elliptical motion similar to the waves caused when a stone is dropped into a pond. It is the slowest, but often the largest and most destructive, of the wave types caused by an earthquake. It is usually felt as a rolling or rocking motion and, in the case of major earthquakes, can be seen as they approach. Named after Lord Rayleigh, the English physicist who predicted its existence.

regional metamorphism: metamorphism resulting from both heat and pressure. It is usually connected with mountain building and occurs over a large area. *See also* contact metamorphism.

reniform: a kidney-shaped mineral habit (e.g., hematite).

reservoir rock: a permeable rock in which petroleum accumulates.

reversed fault: a fault where one slab of the earth's crust rides up over another. Reversed faults are only common during plate collision.

rhyolite: acid, igneous, volcanic rock, typically light in color; volcanic equivalent of granite.

ria: the name for a partly flooded coastal river valley in an area where the landscape is hilly.

Richter Scale: the system used to measure the strength of an earthquake. Developed by Charles Richter, an American, in 1935.

rift, rift valley: long troughs on continents and midocean ridges that are bounded by normal faults.

rifting: the process of crustal stretching that causes blocks of crust to subside, creating rift valleys.

rock: a naturally occurring solid material containing one or more minerals.

rock cycle: the continuous sequence of events that causes mountains to be formed then eroded before being formed again.

rupture: the place over which an earthquake causes rocks to move against one another.

salt dome: a balloon-shaped mass of salt produced by salt being forced upward under pressure.

sandstone: a sedimentary rock composed of cemented sand-sized grains 0.06–2mm in diameter.

scarp slope: the steep slope of a cuesta.

schist: a metamorphic rock characterized by a shiny surface of mica crystals all oriented in the same direction.

scoria: the rough, often foamlike rock that forms on the surface of some lavas.

seamount: a volcano that rises from the seabed.

Secondary Era: an older term for a geological era. Now replaced by Mesozoic Era.

sediment: any solid material that has settled out of suspension in a liquid.

sedimentary rock: a layered clastic rock formed through the deposition of pieces of mineral, rock, animal, or vegetable matter.

segregation: the separation of minerals.

seismic gap: a part of an active fault where there have been no earthquakes in recent times.

seismic wave: a wave generated by an earthquake.

series: the rock layers that correspond to an epoch of time.

shadow zone: the region of the earth that experiences no shocks after an earthquake.

shaft: a vertical tunnel that provides access or ventilation to a mine.

shale: a fine-grained sedimentary rock made of clay minerals with particle sizes smaller than 2 microns.

shield: the ancient and stable core of a tectonic plate. Also called a continental shield.

shield volcano: a volcano with a broad, low-angled cone made entirely from lava.

silica, silicate: silica is silicon dioxide. It is a very common mineral, occurring as quartz, chalcedony, etc. A silicate is any mineral that contains silica.

sill: a tabular, sheetlike body of intrusive igneous rock that has been injected between layers of sedimentary or metamorphic rock.

Silurian, Silurian Period: the name of the third geological period of the Paleozoic Era. It began about 430 and ended about 395 million years ago.

skarn: a mineral deposit formed by the chemical reaction of hot acidic fluids and carbonate rocks.

slag: waste rock material that becomes separated from the metal during smelting.

slate: a low-grade metamorphic rock produced by pressure, in which the clay minerals have arranged themselves parallel to one another.

slaty cleavage: a characteristic pattern found in slates in which the parallel arrangement of clay minerals causes the rock to fracture (cleave) in sheets.

species: a population of animals or plants capable of interbreeding.

spreading boundary: a line where two plates are being pulled away from each other. New crust is formed as molten rock is forced upward into the gap.

stock: a vertical protrusion of a batholith that pushes up closer to the surface.

stratigraphy: the study of the earth's rocks in the context of their history and conditions of formation.

stratovolcano: a tall volcanic mountain made of alternating layers, or strata, of ash and lava.

stratum: (*pl.* strata) a layer of sedimentary rock.

streak: the color of the powder of a mineral produced by rubbing the mineral against a piece of unglazed, white porcelain. Used as a test when identifying minerals.

striation: minute parallel grooves on crystal faces.

strike, direction of: the direction of a bedding plane or fault at right angles to the dip.

Strombolian-type eruption: a kind of volcanic eruption that is explosive enough to send out some volcanic bombs.

subduction: the process of one tectonic plate descending beneath another.

subduction zone: the part of the earth's surface along which one tectonic plate descends into the mantle. It is often shaped in the form of an number of arcs.

sulfides: a group of important ore minerals (e.g., pyrite, galena, and sphalerite) in which sulfur combines with one or more metals.

surface wave: any one of a number of waves such as Love waves or Rayleigh waves that shake the ground surface just after an earthquake. *See also* Love waves and Rayleigh waves.

suture: the junction of 2 or more parts of a skeleton; in cephalopods the junction of a septum with the inner surface of the shell wall. It is very distinctive in ammonoids and used to identify them.

S wave, shear or secondary seismic wave: this kind of wave carries energy through the earth like a rope being shaken. S waves cannot travel through the outer core of the earth because they cannot pass through fluids. *See also* P wave.

syncline: a downfold of rock layers in which the rocks slope up from the bottom of the fold. *See also* anticline.

system: *see* geological system.

tableland: another word for a plateau. *See* plateau.

tectonic plate: one of the great slabs, or plates, of the lithosphere (the earth's crust and part of the earth's upper mantle) that cover the whole of the earth's surface. The earth's plates are separated by zones of volcanic and earthquake activity.

Tertiary, Tertiary Period: the first period of the Cenozoic Era. It began 665 and ended about 1.6 million years ago.

thrust fault: *see* reversed fault.

transcurrent fault: *see* lateral fault.

transform fault: *see* lateral fault.

translucent: a description of a mineral that allows light to penetrate but not pass through.

transparent: a description of a mineral that allows light to pass right through.

trellis drainage pattern: a river drainage system where the trunk river and its tributaries tend to meet at right angles.

trench: *see* ocean trench.

Triassic, Triassic Period: the first period of the Mesozoic era. It lasted from about 225 to 190 million years ago.

triclinic: a crystal system in which crystals have 3 axes, none at right angles or of equal length to one another.

tsunami: a very large wave produced by an underwater earthquake.

tuff: a rock made from volcanic ash.

unconformity: any interruption in the depositional sequence of sedimentary rocks.

valve: in bivalves and brachiopods, one of the separate parts of the shell.

vein: a sheetlike body of mineral matter (e.g., quartz) that cuts across a rock. Veins are often important sources of valuable minerals. Miners call such important veins lodes.

vent: the vertical pipe that allows the passage of magma through the center of a volcano.

vertebrate: an animal with an internal skeleton.

vesicle: a small cavity in a volcanic rock originally created by an air bubble trapped in the molten lava.

viscous, viscosity: sticky, stickiness.

volatile: substances that tend to evaporate or boil off of a liquid.

volcanic: anything from or of a volcano. Volcanic rocks are igneous rocks that cool as they are released at the earth's surface—including those formed underwater; typically have small crystals due to the rapid cooling, e.g., basalt, andesite, and rhyolite.

volcanic bomb: a large piece of magma thrown out of a crater during an eruption, which solidifies as it travels through cool air.

volcanic eruption: an ejection of ash or lava from a volcano.

volcanic glass: lava that has solidified very quickly and has not had time to develop any crystals. Obsidian is a volcanic glass.

volcanic plug: the solidified core of an extinct volcano.

Vulcanian-type eruption: an explosive form of eruption without a tall ash column or pyroclastic flow.

water gap: a gap cut by a superimposed river, which is still occupied by the river.

weather, weathered, weathering: the process of weathering is the mechanical action of ice and the chemical action of rainwater on rock, breaking it down into small pieces that can then be carried away. *See also* chemical weathering and mechanical weathering.

wind gap: a gap cut by a superimposed river, which is no longer occupied by the river.

Set Index

USING THE SET INDEX

This index covers all eight volumes in the *Earth Science* set:

Volume
number   Title
1:    **Minerals**
2:    **Rocks**
3:    **Fossils**
4:    **Earthquakes and volcanoes**
5:    **Plate tectonics**
6:    **Landforms**
7:    **Geological time**
8:    **The earth's resources**

An example entry:

Index entries are listed alphabetically.

plagioclase feldspar **1:** *51*; **2:** 10 *see also* feldspars

Volume numbers are in bold and are followed by page references. Articles on a subject are shown by italic page numbers.

In the example above, "plagioclase feldspar" appears in Volume 1: Minerals on page 51 as a full article and in Volume 2: Rocks on page 10. Many terms also are covered in the GLOSSARY on pages 60–65.

The *see also* refers to another entry where there will be additional relevant information.